Power and Technology

Power and Technology

A Philosophical and Ethical Analysis

Faridun Sattarov

**ROWMAN &
LITTLEFIELD**
──────INTERNATIONAL

London • New York

Published by Rowman & Littlefield International, Ltd.
6 Tinworth Street, London SE11 5AL
www.rowmaninternational.com

Rowman & Littlefield International, Ltd. is an affiliate of
Rowman & Littlefield
4501 Forbes Boulevard, Suite 200, Lanham, Maryland 20706, USA
With additional offices in Boulder, New York, Toronto (Canada), and London (UK)
www.rowman.com

British Library Cataloguing in Publication Information
A catalogue record for this book is available from the British Library

ISBN: HB 978-1-7866-1129-1
ISBN: PB 978-1-7866-1130-7

Library of Congress Cataloging-in-Publication Data

Names: Sattarov, Faridun, 1986– author.
Title: Power and technology : a philosophical and ethical analysis / Faridun Sattarov.
Description: Lanham : Rowman & Littlefield, [2019] | Includes bibliographical references and index.
| Summary: "This book aims to offer an empirically informed philosophical framework for
understanding the technological construction of power, which allows for a differentiated vocabulary for describing various senses of technological power, while bridging together social and
political theory, critical studies of technology, philosophy and ethics of technology"—Provided
by publisher.
Identifiers: LCCN 2019016429 (print) | LCCN 2019980633 (ebook) | ISBN 9781786611291 (cloth) |
ISBN 9781786611307 (paperback) | ISBN 9781786611314 (ebook)
Subjects: LCSH: Power (Philosophy) | Power (Social sciences) | Technology—Philosophy. | Tech-
nology—Political aspects. | Technology—Moral and ethical aspects.
Classification: LCC BD438 .S28 2019 (print) | LCC BD438 (ebook) | DDC 303.301—dc23
LC record available at https://lccn.loc.gov/2019016429
LC ebook record available at https://lccn.loc.gov/201998063

Contents

Acknowledgements

Many people have helped me with their support and encouragement before and during the years I have been researching power and technology. I would like to thank Heather Widdows and Yujin Nagasawa for stimulating my interest in philosophy and global ethics during my graduate years at the University of Birmingham. I am also enormously grateful to Philip Brey, my doctoral supervisor and postdoctoral mentor at the University of Twente. My special thanks go to my colleagues and friends Karen Buchanan, Mark Coeckelbergh, Emiliano Heyns, Anthonie Meijers, Saskia Nagel, Joseph Savirimuthu, and Peter-Paul Verbeek. The book has greatly benefited from the feedback offered by Martin Peterson, Ashley Shew, and Behnam Taebi, and the reviewer who wished to remain anonymous. I would like to express my gratitude to theorists of power Amy Allen, Mark Haugaard, Michael Mann, and Peter Morriss, whose works have inspired the present study. The first part of the book draws on my doctoral research (2010–2015) which was made possible by the Netherlands Organisation for Scientific Research (NWO). I am happy to have had the opportunity of conducting research at some of the finest academic institutions, including the University of Twente, the Eindhoven University of Technology, the University of Liverpool School of Law, the UNESCO Bioethics and Ethics of Science Section, and the 4TU.Centre for Ethics and Technology. The book is dedicated to my first editor Isobel Cowper-Coles.

Introduction

On December 10, 2013, on International Human Rights Day, more than five hundred of the world's leading writers and authors, including five Nobel Prize laureates—Orhan Pamuk, J. M. Coetzee, Elfriede Jelinek, Günter Grass, and Tomas Tranströmer—launched a global petition, condemning the extent of mass surveillance of citizens by states.[1] Specifically, the petition accuses states and corporations of violating the human right *"to remain unobserved and unmolested"* through *"abuse of technological developments"*, while urging the United Nations organisation to create an international bill of digital rights that would ensure the protection of civil liberties in the age of digital technology. This petition is undoubtedly remarkable in many ways, and it remains to be seen whether it can achieve its goal in practice. The petition is noteworthy as it neatly encapsulates a number of recent technological, social, and political developments that have come to characterise the world in the twenty-first century. These developments suggest that the phenomena of technology and power are becoming ever more inseparable. The technological progress of recent decades has presented various individuals, groups, and institutions around the world with novel capabilities to exercise power in ways hitherto unseen. One could even say that technology and power have come to be linked together in such a way that the phrase *"abuse of technological developments"* that figures in the text of the petition could be replaced with the phrase *"abuse of power"* without changing much of the original message of the appeal. Thus, in the context of this petition and beyond, the terms 'technology' and 'power' have become almost synonymous—while technology has come to mean power, power has come to mean technology. *The present book sets out to explore the point of convergence between technology and power by developing an empirically informed philosophical framework for understanding how technology relates to pow-*

1

er, and by laying a provisional groundwork for conceptualising the role of power in the ethics of technology.

BETWEEN TECHNOLOGY AND POWER

The nexus between technology and power has for some time been an object of study in the philosophy, history, and sociology of technology. Lewis Mumford (1964) was perhaps one of the first authors to have claimed that technology can have specific effects on the distribution of power in society. Mumford argued that "from late neolithic times in the Near East, right down to our own day, two technologies have recurrently existed side by side: one authoritarian, the other democratic, the first system-centered, immensely powerful, but inherently unstable, the other man-centered, relatively weak, but resourceful and durable" (Mumford 1964, 2). A decade later, Ivan Illich (1973) drew a similar distinction between "convivial" and "anticonvivial" tools. Convivial tools, which include most hand tools, were defined by Illich as tools that allow for autonomous work and convivial interaction between persons. Anticonvivial tools, which includes most industrial tools, were analysed by Illich as deskilling workers, fostering dependency, and favouring centralised control.

With a nod to Mumford, in his now classic paper titled 'Do Artefacts Have Politics?', Langdon Winner has made the case that "the machines, structures, and systems of modern material culture . . . can embody specific forms of power and authority" (1980, 121). Winner discusses the case of how the structural peculiarities of bridges in New York prevented people who depended on public buses from travelling on the parkways of Long Island thereby exercising power over them (1980, 123–124). Another example cited by Winner refers to mechanical tomato harvesters. Because such harvesters were strongly compatible with large-scale forms of tomato growing, their introduction, in the early 1960s in California, resulted in the concentration of tomato production processes in the hands of large growers, thereby bringing a decline in the number of rural farmers (1980, 126–127). Winner also offers the example of how the operation of nuclear power plants give rise to specific forms and distributions of power: to be operated safely, they necessitate a hierarchical and centralised control, unlike solar energy technology, which can be run in an egalitarian fashion (1980, 130).

The case of the mechanical tomato harvesters discussed by Winner echoes the case study of the automation of the machine-tool industry offered by David Noble (1984). Noble made a valuable input to our understanding of the relation between technology and power by illustrating how automatically controlled tools and machines frequently served the purpose of securing managerial power and control on the factory floor. According to Noble, in addition to their known, or manifest, functions, technological artefacts and systems can have several hidden, or latent, functions that advance the interests and goals of managerial authorities. Noble offered an interesting example of the transformation of the machine-tool industry as it shifted from its reliance on skilled machinists to reliance on automation. Although automated machines were introduced to achieve greater efficiency, the shift to automation also resulted in the deskilling of labour (a process whereby technology is used to replace skilled workers) and the weakening of the labour unions.

Also notable is the work of Richard Sclove to our understanding of technology and power. For Sclove, technologies are "social structures" insofar as they influence social behaviour through various means such as coercive and subconscious compliance, opportunities and constraints, as well as other background conditions (1995, 11–14). Sclove maintains that technologies are "polypotent" entities, meaning that they have potential socio-political outcomes that may not have been intended by their developers or designers (1995, 19–20). Consequently, the "focal function" of a technology (i.e., the intended function for which it was designed) can become outweighed by potentially undemocratic or authoritarian "nonfocal functions" (1995, 21–23). Hence, for Sclove, technological decisions are a kind of political decision. Building on this idea, Sclove argues that citizen participation in technological politics is both morally and politically required. Importantly, technological power should be treated as any other sources of power, such as economic, social, or cultural, because "to neglect any one power medium would be as naïve and imprudent as neglecting another" (1995, 116).

The scholarship on the relation between technology and power has been vastly enriched by the works of Andrew Feenberg, one of the most prolific writers in critical theory and philosophy of technology. In a series of books, including *Critical Theory of Technology* (1991), *Alternative Modernity* (1995), *Questioning Technology* (1999), *Transforming Technology* (2002), and *Technosystems* (2017), Feenberg has developed a critical theory of technology that successfully combines Marxist critique of technology with empirical studies of technology. Feenberg has convincingly argued against both commonsense instrumentalism (the idea that technology is a neutral means to whatever ends chosen by people) and technological determinism (the idea that technological development is constrained by economic or technological rationality; 1999). For Feenberg, technology is an entirely political phenomenon, as it embodies political power, interests and ideologies (2005, 52). Hence, in modern societies, *"masters of technical systems"* wield more power and control over the daily lives and experiences of citizens than *"all the governmental institutions"* (1992, 301).

The connection between technology and power has also become an object of inquiry in sociology and media and communication studies. Most notably, Manuel Castells in a series of works (2010a, 2010b, 2010c, 2007, 2009) devoted to the study of the rise of the network society, offers an illuminating study of the ways in which novel information and communication technologies and practices shape power relations in contemporary network society. Castells observes how following the 1970s there has been a shift from the industrial to the information society, largely as a result of the transformation of the capitalist world order through information technology. While in the industrial society the basic unit of economic organisation was the factory, in the new information society this unit is the network, characterised by instantaneous flows of capital, information, and communication. Consequently, social dependence on these novel modes of exchange and flow of capital and information confers great power to those who control them (Castells 2010a). The media becomes the main political arena (Castells 2009).

More recently, the intuited link between technology and power has received renewed attention in the ethical and critical studies of computer algorithms (e.g., Lash 2007; Beer 2009; Golumbia 2009; Cheney-Lippold 2011; Mager 2012; Snake-Beings 2013; Pasquale 2015; Bucher 2012, 2016; Danaher 2016; Sattarov & Coeckelbergh 2017; Yeung 2017). There is little surprise why there would be heightened interest in algorithmic power, as many would agree that in order to critically and ethically evaluate computer algorithms, it is important first to acquire sufficient understanding of the powers, capacities or capabilities such computer algorithms do actually possess. Moreover, as will be argued below, any adequate ethics or critique of algorithmic practices, cultures, and systems should be grounded on a better, clearer, and more nuanced and differentiated understanding of the many ways in which they can be described as 'powerful'.

In addition to the contributions of the aforementioned authors, there have been a number of other studies concerning the effects of technology on individual and social behaviour that have uncovered phenomena which are closely related to power or which can be regarded as power in some form, but have not explicitly been recognised as such. To concepts that refer to such power-like phenomena, we may include "affordance" (Norman 1988), "script" (Akrich 1992; Latour 1992), "bias" (Friedman & Nissenbaum 1996), "code" (Lessig 2006), and "master switch" (Wu 2010). Certainly not all these concepts refer to precisely the same phenomenon, since they have originated in different theoretical schools and have different scopes assigned to them. However, all these concepts share the idea that technologies can somehow be attributed with various powers to affect and influence their social and physical environment. Similarly, labelling technological artefacts and systems as being 'democratic', 'authoritarian', 'political', 'Western' or 'gendered' (Wajcman 1991) suggests that technologies can embody different forms of power and therefore can have significant social, political, cultural, and economic implications and consequences.

TOWARDS A PLURALIST ANALYSIS OF POWER

Despite the growing interest in the nexus between technology and power, the notion of technological power has not been subjected to robust philosophical treatment. Consequently, little attempt has been made to relate technology to different conceptions of power developed in the social and political theory and philosophy. Most existing discussions on the topic of technology and power often fail to make explicit their conception of power, instead relying on an implicit understanding of power which can often lead to confusion given that authors and their readers may have different, and even conflicting, interpretations of what power means. In contrast, the approach adopted in the book aims to give a systematic account of how technology relates to different conceptions of power extant in the social and political theory and philosophical literature on power. The book aims first to disentangle different senses of power, and then consider how these different senses of power can be applied to technology. The end result should be a more nuanced understanding of technological power appropriately anchored in the contemporary philosophical and social scientific scholarship on power.

Power has had a long intellectual history in philosophy setting up a parade of prominent philosophers and thinkers from Plato through Pareto to Pitkin. The systematic study of the concept of power began in the first half of the twentieth century, most notably with the work of Max Weber (1947). In late 1930s, the eminent English philosopher Bertrand Russell made the case that "the fundamental concept in social science is Power, in the same sense in which Energy is the fundamental concept in physics" (Russell 2004). Less than a century of scholarship later, power has indeed come to be recognised as "the central concept of the social sciences" (Haugaard & Clegg 2009, 1). In this period, power has become of major concern within a variety of fields and disciplines, including philosophy (e.g., Hindess 1996), political science (e.g., Shively 2018), social science (e.g., Russell 2004), social psychology (e.g., French & Raven 1959), international relations theory (e.g., Nye 2011), security studies (e.g., Buzan 2008), history (e.g., Mann 1986), economics (e.g., Bartlett 1989), and so on.

Perhaps partly as a result of its long history, power has been defined variously; for example, as that which *"enable us—or anything else for that matter—to do whatever we are capable of doing"* (Plato 1997, 1104); or, as one's *"present means . . . to obtain some future apparent good"* (Hobbes 1839 [1651], 74); or, as *"the human ability not just to act but to act in concert"* (Arendt 1970, 44). A broad survey of the literature on power can show that the current scholarship abounds with different views and conceptions of power; hence the prevalence of labels and qualifiers such as 'dispositional', 'episodic', 'relational', 'social', 'institutional', 'systemic', 'constitutive', 'Foucauldian', 'Arendtian', and so forth. Given the many views of power, to say that there is only one correct view may lead to dogmatism, whereas saying that all views are correct or incorrect can give rise to relativism or scepticism respectively (Haugaard 2010, 420). Therefore, in our analysis of technological power, we need a pluralist approach to power that can coherently reconcile these different views in a unified framework, while doing justice to different conceptions of power.

The pluralist approach to power adopted in the book maintains that power is a complex multifaceted phenomenon that can be viewed from multiple angles. This approach (presented in detail in Chapter 1) traces its roots to the works of Amy Allen (1999, 2016), Mark Haugaard (2010, 2012), Stewart Clegg (1989), and Sheldon Wolin (2004), among others. In particular, the approach is inspired by Sheldon Wolin, who argued that "concepts like 'power' . . . are not real 'things', although they are intended to point to some significant aspect about . . . things. They represent, instead, an added element, something created by the . . . theorist. Their function is to render . . . facts significant, either for purposes of analysis, criticism, or justification, or a combination of all three" (2004, 6). Thus, power is a conceptual tool that enables us to analyse and critically discuss phenomena or facts—here phenomena and facts concerning technological power.

In the current literature on power, it is possible to identify four main ways of conceiving of power: (1) episodic, (2) dispositional, (3) systemic, and (4) constitutive. Briefly stated, when viewed from the episodic perspective, power is conceptualised as a social relation in which one powerful actor exercises asymmetrical power over another actor, for example, by means of seduction, coercion, manipulation, and so forth. When considered from the dispositional viewpoint, power is regarded as ability, capacity, or potential of an actor or entity to bring about relevant social, moral, or political outcomes. When seen from the systemic standpoint, power is considered to be a property of social, economic, cultural, and political institutions and networks that structurally create possibilities for individual action. Finally, when viewed from the constitutive perspective, power is seen as constituting, or producing, social actors themselves, as if by acting from within these actors. The episodic and dispositional views of power can be described as 'action-centric' views of power, insofar as they focus on the actual or potential actions of persons and groups. The systemic and constitutive views of power, on the other hand, can be described as 'structure-centric' views of power, insofar as these two views regard power as inhering in the social and material context that makes individual action possible. (Chapters 1, 2, 3, 4, and 5 further elaborate on these four views of power.)

The four conceptions of power can be interpreted as corresponding to different levels of analysis that require different levels of abstraction. For example, the dispositional view of power corresponds to micro-level analysis where power is viewed as the properties of individuals; that is, their dispositional properties, such as abilities, capacities, or capabilities. In contrast, the episodic view takes a broader perspective, by considering power as the property of a relationship or interaction between at least two (or more) social actors, given that most episodic views of power are usually formulated in terms of A exercising power over B. The systemic view of power corresponds to macro-level analysis, for it takes an even broader perspective by looking at the various social structures and institutions as structural sources of power. Finally, the constitutive view can be described as combining macro-level and micro-level of analyses with a view to understand or explain how broader systemic and institutional forces and factors can come to constitute individuals. Put differently, a theorist of power either zooms in or zooms out on the picture of power using her lenses for conceptual abstraction.

The particular level of analysis depends on the specific theoretical or normative purposes of the analysis. For example, psychologists may be interested in the episodic or the dispositional views of power, since they may want to understand particular (e.g., seductive, coercive, or manipulative) means that individual actors can adopt as part of their power strategy. Sociologists, on the other hand, may be more interested in large-scale institutions, social and political structures, and technological systems, since they may want to understand how such institutions structure the actions of individuals and persons. We can observe the presence of both action-centric and structure-centric views of power in, for example, Michel Foucault (1980), who being credited for the view of power as ubiquitous, dispersed, and systemic, he does not abandon the level of individual subjects as a level of analysis, by connecting the subject to wider social-institutional structures and processes.

A nuanced understanding of different senses of power can help us understand different senses of technological power. Those interested in studying human-technology relations consisting of individual actors and isolated artefacts can employ the dispositional or episodic views of powers by considering how isolated artefacts can transform individual abilities or capabilities, for example, by empowering or disempowering them. On the other hand, those interested in how large-scale socio-technological systems and institutions can structure social behaviour and action can employ the systemic view of power. However, we must bear in mind that these levels of analyses are not necessarily fragmented or opposed to one another, insofar as particular dispositional properties of artefacts might be stemming from the structural factors of larger socio-technical systems in which they are embedded.

From a different perspective, one inspired by natural sciences, the pluralist approach to power can also be interpreted as a step in the gradual shifting of discourses from more specific theories of power towards a more comprehensive theory of power. To be sure, the present work does not aim to argue against or refute some theory of power. Rather, it proposes to what new and different insights we can gain by considering technology vis-à-vis some of the main conceptions of power as they have been developed in social and political theory. For example, Einstein's physics has not entirely repudiated Newton's physics. The former included the latter into a more comprehensive theory of physics. As Einstein himself expressed it, "The most beautiful fate of a physical theory is to point the way to the establishment of a more inclusive theory, in which it lives as a limiting case" (quoted in Natarajan 2014, 65). From this standpoint, the pluralist approach to power is not the last word on power but a step in the continually evolving scholarship on power.

SITUATING POWER IN THE ETHICS OF TECHNOLOGY

A nuanced understanding of different senses of power can furthermore inform and enhance ethical analyses of technology and technological power. Power is often invoked as something morally bad or reprehensible. In the works of Foucault, however, we hear the call to "cease once and for all to describe the effects of power in negative terms: it "excludes", it "represses", it "censors", it "abstracts", it "conceals". In fact, power produces; it produces reality; it produces domains of objects and rituals of truth" (Foucault 1977, 194). Indeed, on comparison, the episodic view tends to emphasise the strategic means of power such as seduction, coercion, force, manipulation, and so on, as elements of domination (Fay 1987), whereas the dispositional view highlights the constructive side of power, focusing instead on human ability, potential, and capability (Morriss 2002), which are crucial elements of human empowerment and flourishing.

Generally speaking, considerations of power in moral philosophy has a long history. For example, on Aristotle's virtue ethics, the greatest expression of moral virtue requires great political power, insofar as it is the political leader who is in a position to do the greatest amount of good for the community (Kraut 2018). On Kantian deontology, only the good will is inherently good. Power therefore can be good or bad, depending on what use we put it to (Lacewing 2015). Immanuel Kant also held that the possession of power inevitably spoils the free use of reason (Flyvbjerg 1998; Ash 1995). On Mill's utilitarianism (one of the main strands of consequentialism), the only purpose for which power can be rightfully exercised over a person against their will is to prevent harm to others (Mill 1999 [1859]). Similarly, for Jeremy Bentham, another proponent of utilitarianism, coercive power and force is permissible insofar as it promises to prevent some greater evil (1988).

The present work aims to explore the ethics of power and technology in four different directions. First, the four-fold theoretical framework of power (developed in the first part of the book) will be applied to the nascent field of ethics of algorithms with the aim of showcasing how a nuanced understanding of power can shed light on the issues of algorithmic power and bias (see Chapter 6). Second, we shall explore the *conceptual* connection of power to a set (or family) of mainly ethical concepts employed in the ethics of technology. These include the familiar concepts such as responsibility (e.g., of system designers), vulnerability (e.g., privacy vulnerability), authenticity (e.g., of social and moral values), and trust (e.g., in technology, institutions; see Chapter 7). Third, we shall discuss some of the main *practical* implications of power for the processes of making and enacting ethical and socially responsible decisions about technology. In this particular regard, we shall consider some of the main issues and challenges that power presents to the domain of Responsible Research and Innovations, as an increasingly dominant model of decision-making about science and technology in Europe (see Chapter 8). Finally, we shall make a brief foray into recent political economy scholarship to consider those economic and political institutions that play an influential role in determining different paths of technological development in the absence of ethical and democratic structures and frameworks for guiding research and innovation processes in society. Certainly, to do proper justice to the question of power in the ethics of technology would undoubtedly be beyond the scope of the present undertaking. Nonetheless, it is hoped that the described preliminary inquiry can be used as a stepping-stone for future exploration of the many intersections between power, technology, and ethics.

Between power, technology, and ethics, the scope of the present work is vast and may seem impossible. The task presents the author with a dilemma of writing too much and too little. But the point of this project is to develop a sufficiently inclusive theoretical framework, one that can reliably guide our readers through the many issues and topics which frequently intersect in the domains of power and technology. The inclusivity of the book is motivated by an interdisciplinary imperative to show that theories of power, developed within those areas which traditionally have not taken up the question of technology, and which often remain in their separate academic silos, can be brought into philosophical and ethical discussions of technological power. The current scholarship on power is indeed characterised by a number of subfields and subdisciplines with their own unique approaches. While it may not be possible to synthesise all these approaches within the scope of this book, it is worthwhile to attempt to bring into focus, through inevitable abstraction and selection, some of the dominant theories of power and see how they relate to technology. The pluralist approach at least may find some support in Ludwig Wittgenstein, who described his own approach in the following way:

> In teaching you philosophy I'm like a guide showing you how to find your way round London. I have to take you through the city from north to south,

from east to west, from Euston to the embankment and from Piccadilly to the Marble Arch. After I have taken you many journeys through the city, in all sorts of directions, we shall have passed through any given street a number of times—each time traversing the street as part of a different journey. At the end of this you will know London; you will be able to find your way about like a born Londoner. Of course, a good guide will take you through the more important streets more often than he takes you down side streets; a bad guide will do the opposite. In philosophy I'm a rather bad guide. (quoted in McFee 2015, 26)

THE STRUCTURE OF THE BOOK

The present book consists of two parts. Part I (Power and Technology) aims to elucidate different senses in which technology can be described as 'powerful'. With this aim in mind, chapters 1, 2, 3, 4, and 5 develop an integrative philosophical framework for understanding how power relates to the *episodic* (Chapter 2), *dispositional* (Chapter 3), *systemic* (Chapter 4), and *constitutive* (Chapter 5) views of power, which are the four dominant views of power extant in the contemporary literature on social and political theory and philosophy. Part II (Power and the Ethics of Technology) concerns itself with the role of the concept of power in the ethics of technology. Thus, chapters 6, 7, 8, and 9 offer a provisional groundwork for elucidating the role of the concepts of power and technological power in ethical analyses of technology. Chapter 6 applies the theoretical framework developed in the first part of the book to the emerging field of the ethics of algorithms. Chapter 7 explores the conceptual connection of the concept of power to the ethical concepts of responsibility, vulnerability, authenticity, and trust. Chapter 8 considers the practical implications of power for the implementation of Responsible Research and Innovation in Europe. Finally, Chapter 9 considers implications of power to technological innovation in the absence of ethical and democratic institutions and mechanisms for guiding technological innovation processes in society.

Note

1. The text of the petition can be found at: https://www.change.org/p/a-stand-for-democracy-in-thedigital-age-3

Part I

Power and Technology

The overarching aim of the first part of the book is to elucidate different senses in which technology can be said to be 'powerful'. With this aim in mind, the next five chapters (chapters 1, 2, 3, 4, and 5) develop an integrative philosophical framework for understanding how power relates to the *episodic* (Chapter 2), *dispositional* (Chapter 3), *systemic* (Chapter 4), and *constitutive* (Chapter 5) views of power, which are the four dominant views of power extant in the contemporary literature on social and political theory and philosophy.

Chapter One

Four Views of Power

This chapter presents a pluralist approach to the concept of power. On this approach, power is best regarded as a multi-faceted phenomenon that can be viewed from different perspectives. In the current literature on power, it is possible to identify four main views of power: (1) episodic, (2) dispositional, (3) systemic, and (4) constitutive. On the episodic view, power is conceptualised as a relationship in which one actor exercises power over the other, for example, by means of seduction, coercion, manipulation, and so forth. On the dispositional view, power is regarded as a capacity, ability, or potential of a person or entity to bring about relevant social, political, or moral outcomes. On the systemic view, power is understood to be a property of various social, economic, cultural, and political institutions and networks that structurally create possibilities for individual action. Finally, on the constitutive view, power is seen as constituting, or producing, social actors themselves. The first two conceptions (i.e., episodic and dispositional) can be described as 'action-centric' views of power, insofar as they concern actual or potential actions of social actors. The latter two (i.e., systemic and constitutive) can be described as 'structure-centric' views of power, insofar as they regard power as inhering in the social and material context that makes individual action possible. This chapter further elaborates on these four senses of power.

1.1. ISSUES IN DEFINING POWER

Throughout its long history, power has been defined variously, for example, as "a class of the things that . . . enable us—or anything else for that matter—to do whatever we are capable of doing" (Plato 1997, 1104); as being "able to make, or able to receive any change" (Locke 1854 [1689], 359–360); as one's "present means . . . to obtain some future apparent good" (Hobbes 1839

[1651], 74); as "the production of intended effects" (Russel 2004 [1938], 23); as "the probability that one actor within a social relationship will be in a position to carry out his own will despite resistance" (Weber 1947, 152); as "the human ability not just to act but to act in concert" (Arendt 1970, 44); or formulaically as "A has power over B to the extent that he can get B to do something that B would not otherwise do" (Dahl 1957, 202–203). As this multiplicity of definitions suggests, there seems to be an ongoing disagreement about how to define the concept of power. The empirical fact that there is this continuous disagreement is sometimes accounted for by claiming that power is best described as an "essentially contested concept" (e.g., Lukes 2005; Connolly 1993; for a critique of this claim, see, e.g., Wartenberg 1990, 12–17; Haugaard 2010), which is to say that power is one of those concepts which "inevitably involve endless disputes about their proper uses on the part of their users" (Gallie 1956, 169). One of the best-known formulations of this position appears in Steven Lukes (2005), according to whom,

> how we think about [power] relates in a number of ways to what we are trying to understand. Our aim is to represent it in a way that is suited for description and explanation. But our conception of it may result from and be shaped by what we are trying to describe and explain. It may also affect and shape it: how we think of power may serve to reproduce and reinforce power structures and relations, or alternatively it may challenge and subvert them. It may contribute to their continued functioning, or it may unmask their principles of operation, whose effectiveness is increased by their being hidden from view. To the extent that this is so, conceptual and methodological questions are inescapably political and so what 'power' means is 'essentially contested', in the sense that reasonable people, who disagree morally and politically, may agree about the facts but disagree about where power lies. (Lukes 2005, 62–63)

Two theses should be distinguished in the above passage (e.g., Haugaard 2010). The first thesis is the claim that concepts that we use to make sense of our experiences of the world can affect the way in which we see the world. This claim is implied in the sentence "[O]ur conception of it may result from and be shaped by what we are trying to describe and explain". The second thesis is the claim that different conceptions of power can either reveal or conceal the mechanisms and workings of power, whose success or failure can stem from the fact that they are simply hidden from view. When put together, the two theses suggest that "conceptual and methodological questions" concerning power "are inescapably political", and hence power is best understood as an essentially contested concept.

It is worth noting that Lukes is not alone in arguing so. William Connolly, in the *The Terms of Political Discourse* (1993), puts forward similar two theses. First comes the 'conceptual' claim that to describe something is to portray it from a certain perspective, and thus the concepts that we employ in describing reality possess their characteristic features partly because of the perspective from which they are formed (Connolly 1993, 23). This is later followed by the 'political' claim that to argue for one's own position concerning a certain contested concept constitutes a political engagement. For example, when we define 'racism' as 'institutional racism', then we have "shifted the burden of evidence away from the blacks and towards the elites, and thus the balance of political pressures has shifted perceptibly too" insofar as "all those who do not remain (or do not want to be considered) old-fashioned racists" must now show that they are not part of institutional racism as well (Connolly 1993, 202). And again, similar to the argument by Lukes, the two claims together suggest that power is an essentially contested concept.

There is something to be said for the individual premises of the nearly identical arguments of Lukes and Connolly. As far as their conceptual thesis is concerned, it is indeed hard to disagree that the concepts we employ to make sense of phenomena affect the way we perceive those phenomena (e.g., Gray 1978). To use Mark Haugaard's example, someone socialised into indigenous Australian "walkabout time" (in which time is linked to geographical place) is most likely to interpret 'time' differently from a person socialised into modern "linear clock time" (Haugaard 2010, 421). The point Lukes and Connolly make about concepts is not sufficient to make power 'essentially contested', as it would also apply to almost all the concepts that we use. Indeed, the number of concepts that have been claimed to be essentially contested is quite immense to say the least:

> alienation, autonomy, author, bankruptcy, boycott, citizenship, civil rights, coherence, community, competition, the Constitution, corruption, culture, discrimination, diversity, equality, equal protection, freedom, harm, justification, liberalism, merit, motherhood, the national interest, nature, popular sovereignty, pornography, power, privacy, property, proportionality, prosperity, prostitution, public interest, punishment, reasonable expectations, religion, republicanism, rights, sovereignty, speech, sustainable development, and textuality. (Waldron 2002, 149)

As for the political thesis of Lukes and Connolly, there is indeed a political element in disputations concerning how to define power. For example, defining power as 'domination' or 'oppression' is not normally in the interest of the powerful who dominate or oppress others. However, as Haugaard (2010, 421–23) argues a 'contested' concept is not the same as an 'essentially contested' concept. Haugaard invites us to make a parallel with disagreements on the definition of Christianity. Imagine two somewhat fundamentalist Christians. One is a devout Calvinist, and the other is a Catholic. The two may disagree on what should constitute '*the* righteous path' for a Christian to follow. Each of the two will be convinced that that the other is not 'Christian', since the other is not following 'the right' way to salvation. However, if we replace the two fundamentalists with two non-Christian anthropologists, it will not be inherently the case that the two hypothetical anthropologists would have to disagree on their definition of what makes a Christian. Haugaard thus concludes that the non-evaluative usage of the two anthropologists is more appropriate for the purposes of social science.

Haugaard further notes that the claim that power is essentially contested has had the benefit of problematising the notion that there is a single best definition of power. However, the problematisation of the idea of a single best definition of power did not go far enough to demand a pluralist approach to power (Haugaard 2010, 419). Indeed, Lukes's and Connolly's arguments for the essential contestedness of power predict that there will be 'inevitably endless disputes' concerning the definition of power, while simultaneously assuming that there can be a single best definition of power. To illustrate the point, consider how Lukes has argued that power is an essentially contested concept (2005, 62–63), while at the same time making the case that his definition is better than the others (2005, 16, 25, 34, 124). There thus appears to be a contradiction at the very heart of the argument that power is an essentially contested concept, when the singularist thesis that there can be a single best definition of power is accompanied with the essential contestedness thesis that there will be an unavoidably continual disagreement on the definition of power.

Even granting that power is an *essentially* contested concept would not preclude the task of creating a pluralist philosophical framework for understanding the diverse meanings and implications of different ways of conceptualising power. In fact, the essential contestedness thesis should encourage the creation of such a pluralist framework, whereby different views of power can be evaluated, for example, on pragmatic grounds. As Philip Brey notes, "concepts should primarily be evaluated on pragmatic rather than ontological grounds. As Wittgenstein, Peirce and others have shown, we do not usually use concepts to describe objective essences, but rather to selectively highlight aspects of things that are important to us in dealing with them" (Brey 2014, 131). This view of concepts is also true of power, for as Sheldon Wolin has argued: "concepts like 'power' . . . are not real 'things', although they are intended to point to some significant aspect about . . . things. They represent, instead, an added element, something created by the . . . theorist. Their function is to render . . . facts significant, either for purposes of analysis, criticism, or justification, or a combination of all three" (Wolin 2004, 6).

A broad survey of the literature on power can show that the current scholarship abounds with different views and conceptions of power; hence the prevalence of qualifiers such as 'episodic', 'dispositional', 'personal', 'relational', 'social', 'systemic', 'structural', 'institutional', 'facilitative', 'constitutive', 'Parsonian', 'Arendtian', 'Foucauldian', and so forth. In this context, to claim that power is an essentially contested concept would be "either a dereliction of philosophical duty or an admission of failure" (Morriss 2002, 202). Moreover, given the many labels and views of power, to maintain that there can only be one correct definition or concept of power might lead to dogmatism (i.e., this or that definition is correct, while the others are not), or worse still, to scepticism (i.e., all the views are incorrect or irrelevant). Without resorting to outright relativism, as Haugaard (2010) has convincingly made the case, there can be a pluralist approach to power, according to which many existing views and perspectives on power describe a legitimate facet or aspect of power without necessarily being mutually exclusive.

Instead of arguing that there can be a single best definition of power, in the sections to follow an analytical-philosophical approach is adopted, which attempts "to disentangle the different ideas lumped together under one term, and see to what extent the dispute is engendered by the disputants simply talking at crosspurposes" (Morriss 2002, 200–1). As such, the view of power being endorsed can be described as a "plural" (Haugaard 2010, 419), or as an "eclectic" (Castells 2009, 13) view of power, according to which (1) there is a set of concepts each of which, in its own way, describes a relevant aspect of power as a multifaceted social and political phenomenon; and that (2) the relations/differences between these concepts is best conceptualised as being due to their distinct levels of abstraction. The adopted approach promises to remove the dogmatic or sceptical nature of the definitional debates concerning the concept of power, by moving in the direction of an understanding of power, according to which some of the different existing perspectives on power are not necessarily mutually exclusive.

1.2. FOUR VIEWS OF POWER

A majority of existing definitions and concepts of power can be grouped into four main views of power, such as (1) *episodic*, (2) *dispositional*, (3) *systemic*, and (4) *constitutive* views (e.g., Allen 2016; see also Section 1.3, below). Briefly stated, when viewed from the episodic perspective, power is conceptualised as a relationship in which one actor exercises power over the other, for example, by means of seduction, coercion, manipulation, and so on. When viewed from the dispositional perspective, power is regarded as a capacity, ability, or potential of a person or entity to bring about relevant social and political outcomes. When viewed from the systemic perspective, power is understood to be a property of various social, economic, ideological, or political institutions and networks that structurally create possibilities for individual action. Finally, when viewed from the constitutive perspective, power is seen as constituting, or producing, social actors themselves. The first two conceptions (i.e., episodic and dispositional) can be described as 'action-centric' views of power, insofar as they concern actual or potential actions of social actors. The latter two (i.e., systemic and constitutive) can be described as 'structure-centric' views of power, insofar as they regard power as inhering in the social and material context and structures that makes individual action possible. The discussion below elaborates on these four views.

1.2.1. Episodic View of Power

The classic definition of the episodic view of power was offered by Max Weber, who defined power as "the probability that one actor within a social relationship will be in a position to carry out his own will despite resistance" (1947, 152). Weber's classic definition proved to be very influential. In a similar vein, Robert Dahl put forward what he called an "intuitive idea of power", according to which "A has power over B to the extent that he can get B to do something that B would not otherwise do" (Dahl 1957, 202–203). Dahl's account of power triggered a debate in the late 1960s and the early 1970s, which came to be known as the "three-dimensional power" debate (e.g., Lukes 1974). In this debate, Dahl's account was challenged first by Peter Bachrach and Morton Baratz (1962, 1970), and subsequently by Steven Lukes (1974). Nonetheless, although these critics disagreed with Dahl's general account of power, they appeared to have accepted his basic definition of power as an exercise of power-over. As Lukes notes, Dahl's one-dimensional view of power, Bachrach and Baratz's two-dimensional view, and Lukes's own three-dimensional view are all different applications of "the same underlying conception of power, according to which A exercises power over B when A affects B in a manner contrary to B's interests" (Lukes 2005, 30; see also Section 2.1.).

The episodic view of power has also been prominent within social-psychological studies (e.g., French & Raven 1959) and (analytical strand of) critical theory (e.g., Fay 1987; Brey 2008), which are areas of inquiry that have sought to understand the mechanisms and workings of power on the bases of the kinds of incentives and motivations that one actor can offer to another actor within a power relationships, such as reward, coercion, manipulation, persuasion, force, authority, and so on. Although coming from a different philosophical background, Michel Foucault's influential writings on power contain elements which lend support to the episodic view of power, as he claimed that "if we speak of the structures or the mechanisms of power, it is only insofar as we suppose that certain persons exercise power over others" (Foucault 1982, 786). As can be seen, the episodic view of power has been rather influential. While there have been different variations and applications of the episodic view of power, they all seem to share the following prominent features: (1) power is conceptualised in the moment of its actual *exercise* (hence the 'exercise' view of power); (2) power is something that is exercised *over* others (hence the 'power-over' view of power); and (3) power is exercised within social *relations* between two or more actors (hence the 'relational' view power). The episodic view of power and its relation to technology are discussed further in Chapter 2.

1.2.2. Dispositional View of Power

The episodic view of power, however, has been subjected to sustained criticism, most notably by analytically and philosophically inclined theorists, who instead of the episodic view endorsed the dispositional view of power, whereby power is regarded as an ability, capacity, or potential. For instance, Anthony Kenny has argued that

> the capacity of whisky to intoxicate . . . is clearly distinct from its exercise: the whisky possesses the capacity while it is standing harmlessly in the bottle, but it only begins to exercise it after being imbibed. . . . Throughout the history of philosophy there has been a tendency for philosophers—especially scientifically minded philosophers—to attempt to reduce potentialities to actualities. . . . Hume when he said that the distinction between a power and its exercise was wholly frivolous wanted to reduce powers to their exercises. (Kenny 1975, 10)

In this manner, for Kenny, power belongs to the category of dispositional concepts such as ability, capacity, and potential. The dispositional view of power in fact can be traced back to Plato himself, who defined power as what "enable[s] us—or anything else for that matter—to do whatever we are capable of doing" (1997, 1104). Centuries later, Thomas Hobbes similarly defined power as one's "present means . . . to obtain some future apparent Good" (1839 [1651], 74), where one's "present means" are interpreted as one's dispositional qualities. Following suit, John Locke considered power as being "able to make, or able to receive any change" (1854 [1689], 359–360). An earliest modern dispositional definition of power perhaps belongs to Hannah Arendt, for whom power refers to "the human ability not just to act but to act in concert" (1970, 44). While defending the dispositional view of power, Hanna Pitkin wrote of power that "etymologically, it is related to the French *pouvior*, to be able, from the Latin *potere*, to be able. That suggests, in turn, that power is a something—anything—which makes or renders somebody able to do, capable of doing something. Power is capacity, potential, ability, or wherewithal" (Pitkin 1972, 276). It is interesting to note that Steven Lukes, who endorsed the episodic view of power in the first edition of his book *Power: A Radical View* (1974), has embraced the dispositional view of power in the revised second edition of his work, maintaining that power is "a potentiality, not an actuality—indeed a potentiality that may never be actualized" (Lukes 2005, 69; also see Haugaard 2010).

The most comprehensive account of power as a disposition belongs to Peter Morriss, who in *Power: A Philosophical Analysis* (2002) has presented a strong interpretation of power as human ability to bring about relevant outcomes. In his dispositional analysis of power, Morriss draws a distinction between 'ability' and 'ableness' (the latter being a term chosen to denote the noun for able). While 'ability' for Morriss refers to one's capacity to act under certain assumed, or hypothetical, conditions, 'ableness' refers to one's capacity to act when such hypothetical conditions are actualised. To use Morriss's own example, the poor have the ability to eat caviar, while they do not have the ableness to do so (because they cannot afford to buy caviar which is expensive; 2002, 81). Based on this distinction, Morriss argues that what we usually mean by social and political power is a form of ableness, insofar as the study of power in society are concerned with those abilities individuals and groups possess at specific times and places. The distinction Morriss makes between ability and ableness recalls the distinction between 'generic' and 'time-specific' abilities found in the analytical philosophy of action, most notably in the works of Alvin Goldman (1972, 225–226), Honoré (1964, 463), and von Wright (1963, 50–51). The dispositional view of power and how it relates to technology are further explored in Chapter 3.

1.2.3. Systemic View of Power

Whereas the episodic and dispositional views of power focus on the actions and abilities of individuals, the systemic view regards power as the property of broader social, economic, cultural, and political networks, institutions, and structures. On the systemic view, as Haugaard notes, power is regarded as "the ways in which given social systems confer differentials of dispositional power on agents, thus structuring their possibilities for action" (Haugaard 2010, 425). In this way, the systemic view of power underlines how different social, cultural, economic, and political factors and forces "enable some individuals to exercise power over others, or inculcate certain abilities and dispositions in some actors but not in others" (Allen 2016). The systemic view of power, according to Martin Saar (2010), should be interpreted not as a competing alternative to the episodic and dispositional views of power, but rather as a more nuanced and complex view of power, because, on the systemic view, "basic scenario remains individualistic at the methodological level: power operates on individuals as individuals, in the form of a 'bringing to action' or external determination" (2010, 14).

A similar systemic view of power can be found in the theory of structuration developed by Anthony Giddens (1984). Power, according to Giddens, is characterised by what he has described as the "duality of structure": on the one hand, power refers to the capacity of individuals to "make a difference", and, on the other hand, power refers to a structural "property of society or the social community" (1984, 14–15). This view of power corresponds to Giddens's conception of structure. According to Giddens, structure not only constrains individuals but also enables them, for example, through the provision of resources which an individual actor can employ in the course of her interaction with others. That individuals make use of structural resources—resources which can be social and material in nature—suggests that power does not only consist of the actions and abilities of individuals, but also consists of the social and material structural resources. The systemic view of power developed by Giddens therefore serves as an important corrective to the episodic and dispositional views of power which focus primarily on the question of who holds power in society (e.g., Dahl 1961; Morriss 2002).

The systemic view of power is also present in the works of Michael Mann (1986) and Manuel Castells (2009). Mann (1986) focuses on how major ideological, economic, military, and political institutions provide individuals and groups with the resources and means for pursuing their goals. Similarly, Castells (2009) considers how social and communication networks become sources of power on which individual social, economic, and political actors can draw in the course of their interaction with others. These latter two views of power and their relation to technology will be discussed in more detail in Chapter 4.

1.2.4. Constitutive View of Power

The constitutive view of power focuses on the transindividual ways in which individuals are themselves constituted by relations of power (Allen 2016). The constitutive view of power can be traced back to Spinoza, who wrote that "whether a man be wise or ignorant, he is a part of Nature, and everything whereby a man is determined to act should be referred to the power of Nature insofar as this power is expressed through the nature of this or that man" (Spinoza 2002, 683–684). The constitutive conception of power can also be observed in Foucault's formulation of power as "the multiplicity of force relations immanent in the sphere in which they operate and which constitute their own organization; as the processes which, through ceaseless struggles and confrontations, transforms, strengthens, or reverses them; . . . thus forming a chain or system" (Foucault 1978, 92). However, as mentioned in Subsection 1.2.1., Foucault's writings on power also include elements which support the episodic conception of power, when he claims that "if we speak of the structures or the mechanisms of power, it is only insofar as we suppose that certain persons exercise power over others" (Foucault 1982, 786).

As can be seen, the systemic view discussed earlier and the constitutive view appear to be very similar, insofar as they consider power as the property of wider social and historical structures and institutions. However, the difference between the two views is suggested by Amy Allen, when she writes about "*systemic* or *constitutive* conceptions of power—that is, those that view power as systematically structuring possibilities for action, or, more *strongly*, as constituting actors and the social world in which they act" (Allen 2016, emphasis is mine). In this passage we can distinguish between (a) the *weaker* view which conceives of power as "systematically structuring possibilities for action", and (b) the *stronger* view which conceives power as not only creating possibilities for individual action, but also "as constituting actors and the social world in which they act". Allen refers to these weaker and stronger views as "systemic" and "constitutive" respectively (Allen 2016). This distinction between systemic and constitutive conceptions of power is useful in highlighting the difference between, on the one hand, those theorists who primarily focus on how institutions and structures of power provide individual social actors with the means for pursuing their goals (e.g., Mann 1986), and on the other hand, those theorists who go further by attending to how these individual social actors themselves are shaped and constituted by relations of power (e.g., Foucault 1978, 1980, 1982).

1.2.5. Summary of the Four Views of Power

To sum up the discussion in this section, we have identified four main views of power: (1) the episodic view which defines power as an exercise of power over people; (2) the dispositional view which considers power as (human) ability, capacity, potential; (3) the systemic view which regards power as the property of social structures and institutions that create possibilities for individual action; and (4) the constitutive view which conceives of power as constituting individual social actors themselves. The first two views of power (i.e., the episodic and the dispositional) can be described as 'action-centric' view of power, since these views define power in terms of actual and potential actions of social actors. The latter two views of power (i.e., the systemic and the constitutive) in turn can be described as the 'structure-centric' (or 'institutional') views of power, insofar as these views of power consider power as inhering in the social and material structures and institutions that make individual action possible and constitute social actors themselves (see Table 1.1).

1.3. SOME OBJECTIONS CONSIDERED

The pluralist approach to power described in the preceding sections is not without precedent and therefore is not without some defence within social and political theory. A pluralist approach to power that theorises power in terms of several related views, conceptions, or categories of power can be found, for example, in Stewart Clegg, a prominent scholar of power, most notably in his book *Circuits of Power* (1989). In that work, Clegg formulates a theory of power which sees power as a multidimensional phenomenon, consisting of three—episodic, dispositional, and facilitative—levels of power, where changes at one level can have knock-on effects at other levels. Whereas some authors see episodic, dispositional, and facilitative facets of power as disparate or mutually exclusive, Clegg regards them as interrelated. A qualified version of this categorisation of power has been used by Amy Allen (2016), an eminent feminist theorist of power, as a way of carving up the vast literature on power. Allen has thus distinguished between four views of power (i.e., episodic, dispositional, systemic, and constitutive) which also informs the framework described in the preceding sections.

By far the most explicit and sustained defence of the pluralist approach has been offered by Mark Haugaard (2010). Power, according to Haugaard, is a 'family resemblance concept'. The idea of 'family resemblance' was originally developed by Ludwig Wittgenstein (1967), to refer to concepts which share overlapping similarities but without having a common feature that unites them all. Haugaard uses this notion of family resemblance to argue that the different existing conceptions of power, such as episodic, dispositional, systemic, power-to, power-over, and so forth, are all members of a single family of power concepts. This position, according to Haugaard, does not entail relativism whereby 'anything goes', since all concepts of power have to be justified on pragmatic grounds; that is, based on their usefulness as conceptual tools, instead of their ontological essence. Haugaard (2010, 420) thus argues that each of the distinct conceptions of power developed by Dahl (1957), Parsons (1963), Arendt (1970), Foucault (1977), Laclau and Mouffe (1985), Clegg (1989), Žižek (1989), Dowding (1991), Flyvbjerg (1998), Allen (1999), and Lukes (2005) describes a legitimate dimension of power.

Table 1.1.   The Relations between Different Views of Power

Power			
Action-centric view		Structure-centric view	
Episodic view	Dispositional view	Systemic view	Constitutive view

Whereas Haugaard understands the relations between different views and conceptions of power as consisting of family resemblance, it is also possible to interpret these relations in terms of different "levels of abstraction" (Floridi 2008). The four views of power identified in this chapter can be interpreted as corresponding to four different levels of analysis that require different levels of abstraction. For example, the dispositional view of power corresponds to micro-level analysis where power is viewed as the properties of individuals, that is, their dispositional properties, such as abilities, capacities, or capabilities. In contrast, the episodic view takes a slightly broader perspective, by considering power as the property of a relationship or interaction between at least two (or more) social actors, given that most episodic views of power are usually formulated in terms of *A* exercising power over *B* (e.g., Dahl 1957). The systemic view of power corresponds to macro-level analysis, for it takes an even broader perspective by looking at the various social structures and institutions as the constitutive sources of power. Put metaphorically, a hypothetical theorist of power can either zoom in or zoom out on the picture of power using the conceptual lenses for abstraction.

The particular level of analysis depends on the specific theoretical or normative purposes of the analysis. For example, psychologists may be interested in the episodic or the dispositional views of power, since they may want to understand the particular (e.g., coercive, seductive, or manipulative) means that individual actors possess in the course of their social interaction with others. Macro-sociologists, international relations theorists, and political economy scholars, on the other hand, may be more interested in large-scale social, political, or technological institutions, structures, and systems, with a view to understand and to explain how such structural and institutional factors and forces create possibilities for individual or collective actions. Yet, other theorists might find it useful to employ both action-centric and structure-centric perspectives, as can be observed in Foucault's writings on power. Having been credited for the view of power as ubiquitous, dispersed, and systemic, he did not abandon the level of individual subjects as a level of analysis, by connecting the subject to wider social-relational and social-institutional structures and processes.

A sceptic might inquire whether or how the four-fold framework adopted in the present study can accommodate theories of power that may take themselves to be offering a complete account of power. To address this concern, we can note that it is hard to find a theory, at least among the ones considered in this work, that makes a serious and explicit claim to have offered a complete account of power. Even assuming such a theory existed, it should be treated with a healthy dose of scepticism. This point echoes one cautionary anecdote, told by MacIntyre:

There was once a man who aspired to be the author of the general theory of holes. When asked "What kind of hole—holes dug by children in the sand for amusement, holes dug by gardeners to plant lettuce seedlings, tank traps, holes made by roadmakers?" he would reply indignantly that he wished for a general theory that would explain all of these. He rejected ab initio the—as he saw it— pathetically commonsense view that of the digging of different kinds of holes there are quite different kinds of explanations to be given. (MacIntyre 1972, 8)

A complete theory of power would be as unthinkable and unconvincing as a complete theory of holes (Morriss 2002). Furthermore, such complete theories of power may obscure important facets and aspects of power thus creating gaps in our knowledge of them, and "gaps in knowledge, putative and real, have powerful implications, as do the uses that are made of them" (Pasquale 2015, 2). Such gaps in knowledge grant power invisibility thereby affecting the chances of resistance, liberation, and counter-power. There are thus good reasons not to take purportedly grand theories of power at face value.

But what about theories that may offer an account of power that is more expansive than a particular category of the four-fold categorisation? One could argue that Foucault's account of power constitutes one that does not seem to fit entirely into any one category described in the preceding section. For instance, Foucault's thoughts on power, as mentioned earlier, appear to contain claims that support both episodic and constitutive accounts of power. The former can be seen in Foucault's claim that "if we speak of the structures or the mechanisms of power, it is only insofar as we suppose that certain persons exercise power over others" (Foucault 1982, 786). The latter can be seen in his description of power as "the multiplicity of force relations imma- nent in the sphere in which they operate and which constitute their own organization; as the processes which, through ceaseless struggles and con- frontations, transforms, strengthens, or reverses them; . . . thus forming a chain or system" (Foucault 1978, 92). Nevertheless, there is no reason to believe that this would pose a serious issue for the present project, insofar as the aim is not to re-understand Foucault's entire thought on power through the prism of one particular view of power. Rather the aim is to conceptualise and enrich the constitutive dimension of power based on some of the avail- able interpretations of Foucault's thought on power (for more on this see Chapter 5). After all, it is Foucault's works on power that has been a major force for, and a contribution to, the understanding of power that highlights the broader social, institutional, and structural contexts that shape individual relations of power (e.g., Young 1990).

Moreover, the present work does not aim to consider Foucault's thought on power in its entirety, since that would be beyond the scope of a single book. Foucault's writings and commentary on the notion of power is rather vast. As Jon Simons has noted, "[C]ommentary on and critique of Foucault's notion of power has become an intellectual industry in itself" (Simons 1995, 129). The same consideration applies to the general scholarship on power, which stretches beyond the past two millennia. In light of this, we should employ pragmatic conceptual tools to carve up this immense and ever-growing body of literature on power, at least to clarify to ethicists of technology that power can have different meanings within different contexts, and to help students take initial steps on this remarkably diverse intellectual domain we call scholarship on power. The pragmatic categorisation of power is the *first* word; it is by no means the *last*.

1.4. CONCLUSION

To sum up the discussion in this chapter, the current literature on power contains many different definitions of, and approaches to, the concept of power. Faced with the multiplicity of definitions, some authors have claimed that power is an essentially contested concept, meaning that there will always be endless disputes concerning how to define power. The view that power is essentially contested may not withstand criticism. Yet, even assuming that power is an essentially contested concept does not prevent us from treating the concept in a pluralist fashion. Following a number of prominent theorists of power, this chapter has presented a four-fold philosophical framework that categorises power into four main views or conception of power, such as (1) episodic, (2) dispositional, (3) systemic, and (4) constitutive. When viewed from the episodic lens, power is exercised by one actor over another actor. When viewed from the dispositional lens, power is ability, capacity, or potential possessed (but not necessarily exercised) by social actors. When viewed from the systemic lens, power is a property of the overall social system which differentially enables and constrains persons and groups. Finally, when viewed from the constitutive perspective, power is seen as something that constitutes individuals, by often acting from within in a self-imposed manner. The episodic and dispositional views together can be described as 'action-centric' view of power, insofar as they concern actual or potential actions of persons and individuals. The systemic and constitutive views together can be described as 'structure-centric' views of power, since they regard power as inhering in the social, material, or institutional context that makes individual action possible. The following four chapters (i.e., 2, 3, 4, and 5) consider, in more detail, these four views of power and their relations to technology.

Chapter Two

Episodic Power and Technology

This chapter concerns the episodic view of power and its relation to technology. The episodic view (also known as 'power-over') has been prominent within the so-called three-dimensional power debate, as well as within social-psychological and critical social science scholarship that have sought to interpret the mechanisms and workings of power based on the kinds of incentives and motivations that one actor can offer to another actor within a power relationship, such as reward, coercion, manipulation, persuasion, force, and authority. When viewed from the episodic lens, technology can play an important role in the construction, distribution, and exercising of episodic power over others. Based on diverse empirical examples, the discussion in the present chapter aims to illustrate how technological artefacts and systems can play a role in the exercise of episodic power over people, through coercion (e.g., technologies that penalise users for their undesired behaviour), seduction (e.g., technologies that reward users for their desired behaviour), manipulation (e.g., nudging technologies, malicious software), persuasion (e.g., persuasive technologies), force (e.g., physical barriers, handcuffs, shackles), and authority (e.g., guiding technologies, recommendation systems).

2.1. EPISODIC POWER

As noted in Chapter 1, the episodic view of power can be traced back to Weber's definition of power as "the probability that one actor within a social relationship will be in a position to carry out his own will despite resistance" (1947, 152). Weber's classic definition of power seems to have influenced Dahl's definition of power, according to which, "A has power over B to the extent that he can get B to do something that B would not otherwise do" (1957, 202–203). For Dahl, power thus resides in social relations between at least two social actors, in which one actor (whom we can call the power holder) deliberately changes or modifies the behaviour of another actor (whom we can call the power endurer). A social actor can be "individuals, groups, roles, offices, governments, nation-states or other human aggregates" (Dahl 1957, 203).

Dahl's general account of power triggered a debate, which has come to be known as the three-dimensional power debate. The notion of three faces or three dimensions of power (hence the 'three-dimensional power debate') was introduced by Lukes in his book *Power: A Radical View*. In that work, Lukes summarised the debate started by Dahl and his subsequent followers. Briefly stated, the first face of power is associated with the works of Robert Dahl (1961) and Nelson Polsby (1960). The first face of power can be understood as the simple decision-making power. Among the three faces of power, the first face is the most observable, insofar as it concerns the actual behaviour of individuals or groups in decision-making processes. On this view, those people who prevail within decision-making processes are said to have power. The second face of power is associated with the works of Peter Bachrach and Morton Baratz (1962; 1970), who argued that the first face of power overlooks how certain issues never reach the decision-making process. On this view, some persons or groups are powerful, because they have prevailed in the decision-making process simply by controlling the agenda and making certain issues unacceptable for discussion. Bachrach and Baratz call this process non-decision-making. Thus, unlike the first face of power, the second face of power is hard to observe. Finally, the third face has been theorised by Steven Lukes (1974; 2005), who argued that the first and the second faces of power fail to take account of the third face of power that influences the thoughts and preferences of individuals such that they will desire things that they would otherwise oppose. On this view, some issues can be left out of public discussion, simply because people can be influenced not to have preferences for those issues. Unlike the first two faces, the third face of power is even more difficult to observe. Therefore, the third face of power, according to Lukes, is the most insidious form of power.

The episodic view of power dominated this three-dimensional power debate. Although Dahl's approach to power has been challenged (as shown above) first by Bachrach and Baratz, and later by Lukes, even these critics of Dahl's work appear to have agreed with his basic definition of power as an exercise of power over others. As Lukes notes, Dahl's first face of power, Bachrach and Baratz's second face of power, and Lukes's own third face of power have all been different applications of "the same underlying conception of power, according to which A exercises power over B when A affects B in a manner contrary to B's interests" (Lukes 2005, 30). Here we should also note that besides the three-dimensional power debate, the episodic view of power has also been prominent within social psychology research (e.g.,

French & Raven 1959) and within certain areas of critical social science (e.g., Fay 1987). The episodic view has also been used by Philip Brey (2008) in his analysis of the technological construction of social power.

2.2. FORMS OF EPISODIC POWER

Some accounts of the episodic view of power have offered analyses and typologies of various strategic means and incentives that social actors can employ in exercising power over others. Such strategic means usually refer to promises of rewards or threats of punishment, appeals to reason or shared values, uses of force or violence, persuasion, manipulation, and so on. Notably, Bachrach and Baratz (1970, 24–37) put forward a five-fold typology of forms of episodic power: (1) coercion, (2) influence, (3) authority, (4) force, and (5) manipulation. Briefly, in a *coercive* relationship, A secures B's compliance through a threat of punishment (1970, 24). *Influence* involves A getting B to comply "without resorting to either a tacit or an overt threat of severe deprivation" (1970, 30). In the case of *authority*, A secures B's compliance, because B recognises A's authority as reasonable or legitimate (1970, 34). *Force* requires A to impose her will on B despite B's noncompliance, by eliminating B's freedom or possibility of choice (1970, 28). Finally, in *manipulation*, "compliance is forthcoming in the absence of recognition on the complier's part either of the source or the exact nature of the demand upon him" (1970, 28). The following subsections further elaborate on these forms of episodic power.

Bachrach and Baratz's typology of episodic forms of power also appears in Lukes's book *Power: A Radical View* (1974/2005). A qualified version of the typology has also been used by Brian Fay in his *Critical Social Science* (1987). Most notably, the typology has been adopted by Philip Brey in his analysis of the technological construction of social power (2008). However, Brey does not recognise 'influence' as a form of episodic power in his analysis.[1] Instead, Brey includes 'seduction' as an additional form of episodic power, based on the notion of 'reward power' as theorised by social psychologists French and Raven (1959). Our discussion here further qualifies Brey's version of the typology of episodic power one step further, by including 'persuasion' as yet another form of episodic power (see subsections 2.2.6 and 2.3.6).[2] The resulting typology will then consist of six strategic means that social actors can employ in exercising power over others: (1) seduction, (2) coercion, (3) force, (4) manipulation, (5) authority, and (6) persuasion. The following subsections will further elaborate on these specific forms of episodic power.

2.2.1. Seduction

Seduction is a form of episodic power relationship, in which by promise of reward A successfully gets B to do X which B would not otherwise do (e.g., French & Raven 1959, 263). Examples of seduction abound in our daily life. A parent may get her child to clean up his bedroom by promise of a treat. The management of a factory may get increased production results from their employees by promise of promotion or certain other financial incentives. The government of the United States may get the ruling elite of Myanmar to implement progressive democratic reforms by promise of future financial investments in Myanmar. In the area of espionage, so-called 'honey traps' often consist of deploying a sexually appealing operative to seduce someone to give up their secrets or provide information. Seduction as a form of episodic power does not figure in the accounts of power offered by Bachrach and Baratz, it is nonetheless an important means of getting compliant behaviour from others as is evident in social psychology literature. According to French and Raven (1959, 263), in seduction, power stems from the power endurer's perception of the probability that the power holder will provide them with rewards in return for their compliant behaviour. French and Raven further note that the strength of seductive power increases with the magnitude of the reward, which can come either in the form of a provision of positive consequences, or elimination of negative consequences for power endurers by power holders. Since seductive or reward power derives from the probability that the power holder will provide rewards to the power endurer, it follows that successful seduction depends on the power holder's observable behaviour.

2.2.2. Coercion

Coercion is a form of episodic power relationship in which by threat of punishment A successfully gets B to do X which B would not otherwise do (e.g., French & Raven 1959, 263–264; Bachrach & Baratz 1970, 24; Fay 1987, 121). On this definition, coercion involves threats to impose sanctions and penalties, such as causing or inflicting pain or suffering to one's person, a damage to one's property, or a harm to one's loved ones. As Brey (2008, 78) notes, coercion can involve either an addition of new or an increase in existing burdens, all of which are supposed to work against the interest of the power endurer. From the supplied definition of coercion, it can be understood that coercion is a counterpart, or a mirror image, of seduction: whereas seduction is based on the promises of a reward, coercion is based on the threat of a punishment. Just as a promise of promotion can be used as a means of getting compliant behaviour in seduction, so too can a threat of demotion serve as a means of ensuring compliance in coercion. Given this connection between coercion and seduction, French and Raven (1959, 264) note that in some cases it can be difficult to distinguish between coercive and reward power. This in particular applies to cases in which the withholding of a reward can be perceived as a punishment, as well as cases where the withdrawal of punishment is thought of as a reward. We should furthermore note that coercion and seduction can become combined. By combining the words 'threat' and 'offer', Hillel Steiner (Steiner 2007, 288) coined the term "throffer" to refer to an exercise of power that simultaneously involves a coercive threat and a seductive offer. "If you do this for me, I'll give a thousand pounds; if you don't, then I'll beat you up" is an example of throffer.

2.2.3. Force

Force is (arguably) a form of episodic power relationship in which A successfully causes B to do X, by eliminating all of B's options to do anything but X (e.g., Bachrach & Baratz 1970, 28; Fay 1987, 120). An example of force is when a captive person is bound and chained, and thus cannot escape, it means that the captors have eliminated the captive's alternatives to do anything but to remain in the state of being bound and chained. As can be gleaned from this example, force often involves the use of physical force and violence. Nonetheless, there can be instances of force that does not necessarily involve either physical force or violence. Consider, for example, online check-in services used by airline companies, which require passengers to input the details of their travel documents or passports. The web interface of such check-in services can be designed such that they do not allow the input of passports of certain nationalities. The information system here can be described as having a bias against certain kinds of passports (e.g., Friedman & Nissenbaum 1996), whose bearers now must go to and use the check-in desk instead. These hypothetical travellers can be said to have been forced to follow a particular course of action willed by the airline services without recourse to apparent physical force or violence.

Since both force and coercion make references to physical force and violence, the two are often treated as being synonymous. It can indeed be problematic to distinguish between the two notions. For example, Robert Bierstedt (1950) who aims to distinguish between force and coercive power, claims that "force is a manifest [coercive] power. . . . Force . . . means the reduction or limitation or closure or even total elimination of alternatives to the social action of one person or group by another person or group. 'Your money or your life' symbolizes a situation of naked force, the reduction of alternatives to two" (Bierstedt 1950, 733). However, Bierstedt's example of force consisting of a situation where one has to choose between one's life and one's wallet can be interpreted to be an example of coercion. As Bachrach and Baratz argued, in the case of force, A gets what she wants even when B is not complying with the demand, whereas in the case of coercion, A gets what she wants when B complies with the demand (1970, 27). This, according to Bachrach and Baratz, suggests that once A makes recourse to force, B is simply robbed of the choice between whether or not she should comply with A's demand (1970, 27–28). It can thus be argued that force is not a "reduction of alternatives to two" as Bierstedt understood, but a complete elimination of alternatives. Therefore, when B is presented to choose between her life and her wallet, and she chooses her life, it means that A has exercised coercive power over B. On the other hand, if in the same situation B chooses her wallet, and then A has to injure or kill B in order to take the wallet away from B, it means that A has exercised force.

2.2.4. Manipulation

Manipulation, according to Brey, involves "getting people to act in a certain way by performing actions behind their back, withholding information from them, or deceiving them in some way (2008, 82). According to the formal definition of manipulation offered by Fay, "A manipulates B when, by doing x, A causes B to do y [. . .] without B's knowing that A is doing x" (1987, 121). Similarly, for Bachrach and Baratz, the manipulator "seeks to disguise the nature and source of his demands upon" the manipulated, and that the manipulated remains totally unaware that something is being demanded of him" (1970, 31). What these accounts of manipulation share in common is that successful cases of manipulation require that the manipulated should not know that they are being manipulated. Whereas in seduction or coercion, the power endurer is aware of the demands of the power holder, in manipulation the power holder disguises the nature and the source of his or her demands. This we can call the general epistemic condition of manipulation, which can be further specified as follows. First, the manipulated should not know that the manipulator is doing X which is the reason why the manipulated is doing Y. Second, the manipulated should not know that X *is* the reason why the manipulated is doing Y. Third, the manipulated should not know that the manipulator has a reason for the manipulated to do Y. Having thus formulated the specific epistemic conditions of manipulation, we arrive at the formal definition of manipulation offered by Brey:

> *A manipulates B when, by doing x, A causes B to do y which B would other-wise not have done, without B's knowing that (1) A is doing x, or (2) that A is doing x to cause B to do y, or (3) that A has reason r for wanting B to do y.* (Brey 2008, 80)

In their well-known book *Nudge* (2008), Richard Thaler and Cass Sunstein suggest that public policy makers can and should design or arrange context in which people make their decisions such that they promote behavioural changes in the interest of these individuals and society. Thaler and Sunstein discuss a hypothetical case of how the way food is arranged and displayed in a school cafeteria stalls can influence the choice of schoolchildren. Simply by arranging different types of food in certain different ways, it becomes possible to either increase or decrease the consumption of different kinds of food. This form of power and influence Thaler and Sunstein have called a "nudge" (2008, 4).

2.2.5. Authority

Authority is a form of episodic power whereby A successfully gets B to do X by virtue of the fact that B accepts or does not contest the hierarchical or otherwise ranked order in which B is subordinate to A (or where A is super-ordinate to B). One can find instances of authority in cases when the taxpayer complies with the demands of the tax collector, or when the vassal obeys the will of the lord, or when the churchgoer obeys the word of the priest. Here authority should be distinguished from coercion. For example, it can be the case that A gets B to comply with A's demands, but where B's compliance is a result of B's fears of facing potential reprisals if she does not comply, and not the result of B's acceptance of her subordinate position. Authority in its pure form thus requires that B's compliance should derive from B's acceptance of the state of affairs in which B is subordinate to A.

Max Weber famously identified three different types of authority: (1) *legal*, (2) *traditional*, and (3) *charismatic* (1947, 324–363). *Legal authority* as a form of power is held by a person who occupies a lawfully established superordinate position. It can be argued that the legal authority does not really belong to the person who exercises it, but, more appropriately, to the official role or status which the person occupies. For example, when one obeys a police officer, one can be said to be obeying the *officer* and not really the *person* who happens to be this officer. This furthermore suggests that legal authority can be exercised by an institution or organisation. *Traditional authority* as a form of power is usually held by a person who occupies her authoritative position in accordance with a certain tradition, often on the bases of their age, sex, caste, or some other such characteristic. On this view, gerontocracy, patriarchalism, and patrimonialism are all instances of superordination and subordination based on traditional authority. Finally, *charismatic authority* as a form of power is held by persons who are perceived to possess certain outstanding or exemplary virtues inaccessible to common people. Prophets, wartime leaders, and revolutionaries often happen to be charismatic leaders.

2.2.6. Persuasion

Persuasion is a form of episodic power, whereby A gets B to do X by getting B to accept that it is reasonable for B to do X. Persuasion normally involves a person or a group convincing another person or group to assent to a certain course of action or to accept a certain state of affairs. The power to persuade is the power to convince others with the help of credible but not necessarily proven evidence (e.g., Uhr 2011, 478). On this view, persuasion does neither require a full disclosure on the part of the persuader nor a fully informed consent on the part of the persuadee. Persuasion occurs at different levels of social and political interaction. While persuasion is normally regarded as an interactive process happening between two or more persons or groups, it is also possible to conceive of persuasion as an internal process of deliberation taking place within one person or group, for example, when we try to persuade and convince ourselves regarding some matter.

Persuasion is one of the classic forms of episodic power, as evidenced in the history of ancient rhetoric (e.g., Takács 2008; Garsten 2009). Some of the oldest examples of rhetorical persuasion extant in the world literature come from the *Iliad* attributed to the legendary Greek poet Homer. For instance, the *Iliad* features Paris persuading Helen of Sparta, to leave her husband and elope with him, with this elopement eventually leading to the Trojan War (e.g., Smith 2017, 201). Book IX of the *Iliad* contains the narrative of how the warrior Achilles is persuaded by some of his friends to rejoin the Greeks in the battle against Troy. The friends make their case by appealing to honour, obligation, and interest in their attempt to get Achilles to rejoin their military campaign. This example, as Uhr (2011, 478) notes, contains the standard list of means of strategies of persuasion and has become an example that heavily frequently figures in the scholarly literature on persuasion.

Persuasion, as a form of episodic power, plays an important role within democratic political practices. For example, political parties and political leaders deploy persuasive tactics in building and maintaining their political base. Episodic forms of power such as coercion and force can be insufficient or inappropriate for political leaders in processes of pursuing 'public relations' goals, such as maintaining trusting attitudes with the followers or winning over the electorate. Hence, successful politicians are those leaders who possess the skills, qualities, and experience which make them persuasive on political campaign trails and within public deliberations forums. In the context of U.S. politics, "presidential power is the power to persuade" argues Richard Neustadt in his influential book *Presidential Power* (1991, 11). Persuasion also plays an important role at the international relations level. At this level, persuasion shares a similarity to what international relations theorists call soft power (e.g., Nye 2011; Rothman 2011).

2.3. EPISODIC POWER AND TECHNOLOGY

Having presented six different forms of episodic power in the previous section, the present section aims to highlight the ways in which technology plays a role in creating and sustaining episodic relationships of power. In doing so, the discussion here will draw on the analysis of the role of technology in the construction of episodic power offered by Brey (2008). Technology, as theorised by Brey, can play three different roles in creating and maintaining relations of episodic power (which Brey calls 'social power'):

> (1) Technology plays a *proximate* role in the exercise of episodic power, when A and B are at the same location, where A uses technology T to do X in order to get B to do Y. An example of this would be a police officer armed with a pistol and a baton, where the pistol and the baton grant the police officer more effective power to perform the actions necessary for coercing or forcing a possible suspect. Since such cases of episodic power require that both the power holder and power endurer should be at the same location, Brey calls such cases 'proximate' relations of power.

> (2) Technology plays a *distanciated* role in the exercise of episodic power, when A and B are at different locations, and A uses technology T at B's location to do X in order to get B to do Y. An example of this would be a traffic police officer sitting at her office, while using a remote-controlled speed camera to observe speeding drivers and thus modifying these drivers' behaviour from a distance. Since such cases of episodic power involve the power holder exercising power over the power endurer from a distance, such cases can be called 'distanciated' relations of power.[3]

> (3) Technology plays a *delegated* role in the exercise of episodic power, when A and B are at different locations, and A ensures that technology T is at B's location to do X in order to get B to do Y. An example of such exercises of power is a code-controlled access door that can only let in those who know the correct access code; or a speed bump that can slow down traffic. Since such cases of episodic power do not require the power holder to engage with the technology in either proximate or distanciated way, following Latour (1992), Brey calls them 'delegated' exercises of power (Brey 2008).

On further reflection, the above analysis of the role of technology in the exercise of episodic power seems to be compatible with the so-called *principle of generalised symmetry*, according to which any elements within an analysis of social phenomena, whether they are social, natural, technical, or material, should be attributed the same explanatory role and should be analysed using symmetrical vocabulary (e.g., Law 2012, 124). The principle of generalised symmetry suggests that both human actors and nonhuman artefacts can be delegated to perform certain actions on behalf of others—that is, there is a symmetry between humans and artefacts in performing those actions (e.g., Latour 1992). On this view, a social actor wishing to enhance her episodic power over others can employ other humans and/or artefacts to achieve her objective. In a proximate exercise of episodic power, the actor can bring along some*one* or some*thing* to the scene of the exercise of power; in a distanciated exercise of power, the actor can dispatch a proxy agent or artefact to the scene, while instructing or controlling them from a distance; and, in a delegated exercise of power, the actor can dispatch an agent or an artefact to the scene, having instructed or programmed them in advance.

The use of technology in proximate and distanciated exercises of episodic power is important, yet it is the delegated exercise of episodic power that is more interesting, since in the delegated scenario, the power holder actor is absent from the scene of the exercise of power, and technologies are therefore perceived as yielding power over others. For this reason, the remainder of this section will mainly concern the role of technology within a delegated exercise of power. In any case, whether proximate, distanciated or delegated, the exercise of power over others happens through the same processes identified in Section 2.2., processes such as seduction, coercion, force, manipulation, authority, and persuasion.

2.3.1. Seduction and Technology

Delegated seduction occurs when a technological artefact or system is designed such that it can deliver positive consequences (or remove negative consequences) for their users by requiring these users to perform (or refrain from performing) certain actions and, in this manner, cause these users to do a certain action which these users would not otherwise do. An example of delegated seduction is when a vending machine in a remote corner of a vast airport asks an unusually high price for a packet of crisps or a bottle of water. In this example, seduction is distinguished from cases of fair exchange or purchase of goods or services. What can distinguish the former from the latter is that seduction as an exercise of episodic power involves an asymmetrical relation of power whereby a powerholder delivers positive consequences to the power endurer as a result of unfair or inequitable exchange. On this view, a vending machine that asks a fair price for a product it sells would not constitute a case of delegated seduction, whereas the vending machine in the earlier example would (for a similar point see Brey 2008, 84).

Clickbait, as its name suggests, is a website link designed to seduce, or to 'bait', users to visit some other website. So-called 'pop-up' web advertisements offering barely credible products and services (which often range from miracle cures to promises of teaching a foreign language in a matter of hours) are also often designed to seduce internet users to give up sensitive personal information about themselves. Since these instances of delegated seduction frequently involve deception, they should be better understood as a hybrid form of episodic power that combines elements of seduction and manipulation. Another example of online seduction is when social networking websites and domains, such as Facebook and Gmail, offer their users 'free' services, where such 'free' services are in fact based on users' giving up their personal information which can then be used by these internet firms for the future purposes of targeted advertising.

2.3.2. Coercion and Technology

Delegated coercion occurs when a technological artefact or system is designed such that it can deliver negative consequences to (or remove positive consequences from) users and other people in their environment, and, in this manner, cause these users and persons to perform actions which they would otherwise not perform. Brey (2008, 83) cites several cases of delegated coercion. For instance, speed bumps have an effect of preventing drivers from speeding, given the fact that driving over a speed bump at an accelerated speed normally causes a strong physical impact that can be very harmful to their vehicles. Similarly, speed cameras have an effect of slowing down drivers, when the number plates of those vehicles that are violating regulated speed limits are photographed and subsequently used by law enforcement agencies to penalise their owners or drivers. However, we should note that there is an important difference between speed bumps and speed cameras: in the case of speed bumps, the capacity to penalise is contained within the artefact itself, whereas in the case of speed cameras, the capacity to penalise is distributed across a socio-technical system which comprises both the speed camera and the penalising authorities. Taking this difference into account, Brey suggests drawing a distinction between *intrinsic* and *derived* powers of technological artefacts and systems, with speed bumps having intrinsic powers and speed cameras having derived powers. The distributed character of speed cameras further suggest that their successful operation depends on other human and non-human actants within the distributed system.

Other examples of delegated coercion popularised by Bruno Latour (1992) include a hotel key with a heavy key ring, which coerces hotel guests to return the key to the hotel reception before they leave the hotel, and an automobile seatbelt system that emits high-pitched repetitive signals if the driver starts the engine without putting on his or her seat belt. However, we should note that not all negative consequences are the result of delegated coercion. For example, one can receive an electric shock as a result of using an electrical appliance or equipment in an unsafe manner. In this example, the negative consequence is not intended by some distant social actor (e.g., the designer or the manufacturer) but is an unintended outcome resulting from the incorrect use of the artefact or system in question.

2.3.3. Force and Technology

Delegated force occurs when a technological artefact or system is designed such that it can force, forbid, or constrain the physical movements of users or persons who may be in their environment. Forcing and constraining capacity obviously is something intrinsic to physical and material artefacts by virtue of their physical structure and mass. Examples of delegated force include artefacts such as handcuffs, shackles, straitjackets, chastity belts, and muzzles which can be used to physically prevent a person from performing certain actions or behaviours. We are in fact often surrounded by walls and fences, which are cases of delegated force to constrain or direct our movements, or to prevent us from entering or leaving certain enclosed areas. Artefacts, as Brey (2008, 84–85) notes, can also be designed to constrain human perception and communication. For example, a car consisting of two separated compartments can prevent the chauffeur and the passengers from talking or seeing one another. Similarly, a room with no windows prevents people from looking outside. To use the famous example discussed by Winner (1980), low-hanging overpasses over highways can prevent larger vehicles from driving under them. Indeed, as David Norman has argued, many technological artefacts possess forcing functions; that is, physical constraints requiring users to perform certain actions indirectly related to the purpose to which they want to use the artefact (Norman 1988, 132–138, 204–206).

However, there can also be cases of delegated force that are subtler than handcuffs, shackles, and walls. Consider, for example, items of clothing and footwear, such as corsets, brassieres, and shoes with high heels. Such items are often purposefully designed to constrain the bodily movements of (mostly) women who wear them. As such, they represent an interesting case of delegated force, insofar as they are self-imposed exercise of power. Some theorists might be tempted to rule out such cases as being an exercise of power, since they are self-imposed. Nonetheless, as we shall see in Chapter 5, self-imposition can be a result of constitutive power relations and influenced by dominant patriarchal power structures.

2.3.4. Manipulation and Technology

Delegated manipulation occurs when a technological artefact or system is designed or arranged such that it can cause people to perform certain actions or behave in specified ways, yet without making these people aware that they are being influenced in this manner. In Section 2.2.4., we have already considered an example of manipulation, in which different foods in a school cafeteria arranged and displayed in such a way so as to influence the choice of school children. Although this example fits the definition of manipulation, Thaler and Sunstein (2008) prefer the term 'nudge' instead of 'manipulation', given that such cases of behavioural influence do not harm the interests of people, or in the above example, those of schoolchildren. Brey (2008, 85) also cites similar examples of harmless manipulation, or nudging. For example, the items of furniture in the interior of a room can be arranged such that the arrangement forbids or encourages people to behave in certain ways. Similarly, tables and chairs can be arranged in order to highlight the differences in status among people.

Nonetheless, there are also cases of more sinister and harmful forms of technological manipulation. Consider, for example, a recent case study, published in the *Proceedings of the National Academy of the Sciences* (PNAS), presenting "experimental evidence for massive-scale contagion via social networks" by tweaking the number of positive or negative posts shown to Facebook users by experimentally manipulating users' algorithmically managed 'News Feed' (Kramer et al. 2014). This algorithmic feature of Facebook makes 'decisions' regarding which of the other users' 'status updates' or news items Facebook users should see on their home page. The study claimed to show that such news and updates can influence the mood and feelings of Facebook users as well as the character of their subsequent posts. There are also even more deceitful instances of technological manipulation involving malware and spyware (i.e., malicious computer software specifically programmed to perform harmful operations on computer on which they become installed). Once installed, such malware and spyware can take partial control of the computer's operation, monitor the user's computing activity, alter the computer's settings, install additional software, and the like. Malicious programmes of this kind fit the definition of delegated manipulation— the programme says it does one thing, while in fact it does something completely different behind the back of the user.

2.3.5. Authority and Technology

Authority as a form of power invariably involves delegation, whether human or technological. This is to say that even in its non-technological instances of delegation, authority involves an act of delegation of some other higher authority or power. To illustrate the point, consider some of the theories of state authority. Whether it is the doctrine of the Divine Right of Kings (e.g., Figgis 1914) or some version of the social contract theory (e.g., Hobbes 1839), the authority of the state involves some form of delegation: in the former case, the sovereign authority is ordained by God, while in the latter case, the state authority is the outcome of collective delegation of powers by individuals (see also Section 5.2). On these views, authority is a delegation of some other authority.

While authority is traditionally delegated to humans, there are cases of technological delegation of authority, where artefacts are designed such that they express the directives of commands of some other remote authority. In cases of delegated authority, it is "symbolic means rather than physical force" that brings about a certain required behaviour in people (Brey 2008, 86). For example, 'no parking' sign on the side of a street is an artefact delegated with the authority of traffic warden or police, which through symbolic means expresses the directive 'not to park'. Those who recognise the authority of the agent responsible for placing the 'no parking' sign, would not park their vehicles there. The same applies to other traffic signs.

2.3.6. Persuasion and Technology

Delegated persuasion takes place when a technological artefact or system is designed or arranged such that it communicates symbols or information by employing principles of persuasion (such as credibility, trust) with the aim of changing the behaviour or attitude of persons. Just like delegated authority, delegated persuasion relies on symbolic means than on physical force in inducing persons and users to behave in desired ways. Mundane artefacts such as leaflets, notice stands, instruction manuals, are all examples of persuasive technologies.

Brian Fogg (2003) has coined the term captology to refer to the study of persuasive technologies. According to Fogg (2003), there are five different principles that can be employed in persuading people via interactive media technologies: *attractiveness*, *similarity*, *praise*, *reciprocity*, and *authority*. The principle of *attractiveness* holds that a persuasive technology that is visually attractive to its users is likely to be more persuasive. The principle of *similarity* states that people are more easily persuaded by artefacts that are, in certain ways, similar to themselves. The principle of *praise* maintains that interactive technologies can lead users to be more open to persuasion, by expressing praise through symbols, words, images, or sounds. On the principle of *reciprocity*, persons and users can feel the need to reciprocate, when a persuasive technology has done them a favour. Finally, according to the principle of *authority*, technologies that assume roles of authority will have enhanced powers of persuasion.

According to some studies in the area of captology, persuasive technologies with anthropomorphic features can more effectively induce behavioural change in users when compared to technologies without such features (e.g., Pak et al. 2012; Fogg 2003). These findings also suggest that developers of such persuasive technologies can exert greater influence and asymmetrical power over technology users, while raising important ethical questions about trust and justification of the motives and goals of those behind the persuasive technology (Berdichevsky & Neuenschwander 1999). For this reason, persuasive technologies are being studied from the ethical perspective. For instance, Spahn (2012) has argued that ethical guidelines for the development and usage of persuasive technologies can be derived from applying discourse ethics (e.g., Habermas 1996) to this type of technologies.

2.4. CONCLUSION

To summarise the discussion in this chapter, the episodic view of power has been prominent within the three-dimensional power debate that took place in the 1960s and 1970s, as well as in social psychology and critical social science scholarship which have sought to interpret the mechanisms and workings of power based on the kinds of incentives and motivations that one actor can offer to another actor within a power relationship, such as reward, coercion, manipulation, and so forth. On this view, episodic power relations involve one actor exercising asymmetrical power over another actor. Here a social actor can be a person, a group, an organisation, or an institution, such as the state, the church, the army, and so on. In exercising episodic power, these social and political actors can make use of various strategic means in order to get what they want, means including seduction, coercion, force, manipulation, persuasion, or authority. A social actor may choose this or that strategic means, depending on the specific nature of the social relationship and the particular circumstances in which power is to be exercised. Importantly, technology can play a delegated role in creating and maintaining episodic power relations. An example of such exercises of power is a code-controlled access door that can only let in those who know the correct access code; or a speed bump that can slow down traffic. Such cases of episodic power do not require the power holder to engage with the power endurer directly, by leaving or delegating technology to do the job.

Notes

1. For a good discussion of conceptual differences between influence and power, see also Morriss (2002, 29–32).

2. To be sure, Brey (2008) discusses 'persuasion' and its relation to technology in his analysis. However, in his analysis 'persuasion' is subsumed under the cateogry of 'authority'.

3. Brey (2008, 82) describes this as "distal" exercise of power. However, it is better to describe this kind of technological exercise of power as "distanciated", since "distal" as a term is more prevalent in anatomy than sociology (see also Giddens 1991, 14).

Chapter Three

Dispositional Power and Technology

This chapter focuses on the dispositional view of power and its relations to technology. The dispositional perspective of power (also known as 'power-to') regards power as ability, capacity, or potential and has been largely advocated by analytically and philosophically inclined social and political theorists, who endorsed the dispositional view of power as a better alternative to the episodic view of power discussed in the preceding chapter. When viewed from the dispositional lens, technology can be attributed dispositional properties or capacities that can bring about outcomes of social, moral, and political relevance. Depending on their degree of autonomy, technological dispositions and capacities can become an important constitutive part of human abilities and capabilities. While there are some important differences between human abilities and technological capacities, in our increasingly technological age we may need novel concepts for describing the powerful dispositional qualities of technological artefacts and systems, especially given that the effects and consequences of certain kinds of modern technologies cannot be easily traced back to the designers of these technologies.

3.1. DISPOSITIONAL POWER

The discussion of the dispositional view of power in this chapter closely follows the dispositional analysis of power offered by Morriss (2002), whose analysis in turn draws on the works of Gilbert Ryle (2009), Anthony Kenny (1975), Aristotle (*Theta* 1046a–1048a), John Austin (1961), and Alvin Goldman (1972) in some key respects. His analysis of the dispositional aspect of power begins by alerting us to the distinction between dispositional and episodic terms and concepts popularised by Gilbert Ryle in *The Concept of Mind*. According to Ryle, "to say that a person knows something, or aspires to be something, is not to say that he is at a particular moment in process of doing or undergoing anything, but that he is able to do certain things, when the need arises, or that he is prone to do and feel certain things in situations of certain sorts" (Ryle 2009, 100). For Ryle, verbs such as "to know", "to aspire", and "to possess" are dispositional terms that describe regular tendencies of human beings to act and behave in certain ways, which is different from episodic terms that describe events, happenings, and occurrences. The difference between dispositional and episodic concepts reflects the distinction we usually make between 'having power' and 'exercising power' (see also Oppenheim 1961, 100).

Given that episodic and dispositional concepts refer to rather distinct phenomena, they should by no means be confused. However, according to Kenny (1975) there exist two related fallacies frequently committed in interpreting the episodic and dispositional properties of the observable world. On the one hand, an existence of a disposition has sometimes been confused with its exercise (which is called 'the exercise fallacy'); on the other hand, an existence of a disposition has often been confused with its vehicle (hence 'the vehicle fallacy'). The two fallacies are well explained by Kenny:

> Consider the capacity of whisky to intoxicate. The possession of this capacity is clearly distinct from its exercise: the whisky possesses the capacity while it is standing harmlessly in the bottle, but it only begins to exercise it after being imbibed. The vehicle of this capacity to intoxicate is the alcohol that the whiskey contains: it is the ingredient in virtue of which the whisky has the power to intoxicate. The vehicle of a power need not be a substantial ingredient like alcohol which can be physically separated from the possessor of the power. The connection between the power and its vehicle may be a necessary or a contingent one. It is a contingent matter, discovered by experiment, that alcohol is the vehicle of intoxication; but it is a conceptual truth that a round peg has the power to fit into a round hole. Throughout the history of philosophy there has been a tendency for philosophers—especially scientifically-minded philosophers—to attempt to reduce potentialities to actualities. But there have been two different forms of reductionism, often combined and often confused, depending on whether the attempt was to reduce a power to its exercise or to its vehicle. Hume when he said that the distinction between a power and its exercise was wholly frivolous wanted to reduce powers to their exercises. Descartes when he attempted to identify all the powers of bodies with their geometrical properties wanted to reduce powers to their vehicles. (Kenny 1975, 10)

There are thus good reasons to regard power as some sort of a disposition. Upon closer examination, there appears to be more than one kind of dispositions. Consider, for instance, the distinction drawn by Aristotle

between rational powers, such as the ability to speak Greek, and natural powers like the power of fire to burn. If all the necessary conditions for the
exercise of a natural power were present, then, he maintained, the power was
necessarily exercised: put the wood, appropriately dry, on the fire, and the fire
will burn it; there are no two ways about it. Rational powers, however, are
essentially, he argued, two-way powers, powers which can be exercised at
will: a rational agent, presented with all the necessary external conditions for
exercising a power, may choose not to do so. (Kenny 1975, 52–53, citing
Aristotle, *Theta* 1046a–1048a)

The distinction drawn by Aristotle reveals that there are "ordinary dispositional powers" and dispositional powers "which can be exercised at will"
(Morriss 2002, 25). Our disposition to cry when one is pepper-sprayed in the
face is an example of the former, whereas one's dispositional power to write
a letter to a friend whenever one wants to do so is an example of the latter.
The former kind of dispositions can be called reflexive dispositional powers,
while the latter kind of dispositions can more appropriately be called abilities
since abilities are dispositional qualities with an intentional component (e.g.,
Maier 2018). On this view, dispositional powers when attributed to human
agents should be understood as abilities.

In ascribing abilities to people, a number of noted philosophers, among
them Honoré (1964), von Wright (1963), Goldman (1972), and Morriss
(2002) distinguish between *generic* and *specific* abilities. Generic abilities
are abilities that people have despite their particular circumstances, whereas
specific abilities are those the possession of which depends on the particular
circumstances. To get a better grasp of the distinction, consider the following
passage from John Austin:

> We are tempted to say that 'He can' sometimes means just that he has the
> ability, with nothing said about opportunity, sometimes just that he has the
> chance, with nothing said about ability, sometimes, however, that he really
> actually fully can here and now, having both ability and opportunity. . . . The
> only point of which I feel certain is that . . . can . . . [has] an all-in, paradigm
> use, around which cluster and from which divagate, little by little and along
> different paths, a whole series of other uses, for many of which, though per-
> haps not for all, a synonymous expression ('opportunity', 'realize', and so on)
> can be found. (Austin 1961, 178)

As Austin noted, there is more than one sense of 'can', for consider the
following two sentences: (1) George can read (only if he has got his glasses),
and (2) George can read (now that he has got his glasses). As can be seen, the
'can' in the first sentence refers to George's ability to read in general, the
ability that stays with George whether he has got his glasses or not, which is
his ability that sets him apart from illiterate people. However, the 'can' in the
second sentence refers not only to George's general ability to read but also to

the fact that he has the means to read here and now. Following Austin, the 'can' in the first sentence can be called 'can of ability', while the 'can' in the second sentence can be called 'can of opportunity'.

In fact, several writers on ability and power, most notably von Wright, Honoré, Goldman, Morriss, Mele, and Maier have distinguished between two senses of ability, which correspond to the two senses of 'can' distinguished by Austin. For von Wright this is the distinction between what he called "can do of ability" and "can do of success" (1963, 50–1); for Honoré this is the distinction between "general can" and "particular can" (1964, 463); for Goldman this is the distinction between abilities and time-specific abilities (1972, 225–226); for Morriss this is the distinction between "generic ability" and "ableness" (2002, 48–49); for Mele it is the distinction between "general practical ability" and "specific practical ability" (2003, 447); for Maier it is the distinction between "general abilities" and "specific abilities" (2018) (see table 3.1 below).

Generic abilities refer to something that is the property of the power or ability holder and not of the environment: the claim is about what the individual can do assuming the individual has the opportunity. As such, generic abilities are often qualified by adverbs such as 'usually', 'normally', 'invariably', 'occasionally', and the like (e.g., Honoré 1964, 465). In contrast, specific abilities refer to something that is both the property of the individual and the environment; the claim is about whether the power holder has both the ability and the opportunity to do something. As such, specific abilities are used with specific times or places attached, and thus are often expressed through 'was/is/will be able to' (e.g., Morriss 2002, 82). The following passage from Morriss better clarifies the distinction:

> The rich are able to feed off caviar and champagne; the poor have to restrict themselves to beer and pickles, and are unable to eat more expensive food. This is not because of any lack of masticatory ability on their part, but because

Table 3.1.   Generic and Specific Abilities

	I. Generic	II. Specific
Austin	can of ability	can of opportunity
von Wright	can do of ability	can do of success
Honoré	general can	particular can
Goldman	ability	time-specific ability
Morriss	generic ability	ableness
Mele	general practical ability	specific practical ability
Maier	general abilities	specific abilities

of the social and economic environment they inhabit. They are unable to eat caviar, whilst having the ability to do so. (Morriss 2002, 81)

Thus, generic abilities can be understood as being in some sense prior to specific ability—that is, to possess a specific ability is to possess a generic ability and to meet some further constraint, such as having an opportunity. This kind of interpretation appears to be in agreement with a view arguably implicit in what have been called the "new dispositionalist" approaches to ability (Maier 2018). What is noteworthy here is that the move from a generic ability to specific ability can be described as a gradual process. Consider, for example, a well-known historical event—the assassination of Franz Ferdinand, the archduke of Austria-Hungary, by Gavrilo Princip in Sarajevo in June 1914—which is widely believed to have ultimately led to the outbreak of the World War I. Before the assassination actually took place, we can assume that Gavrilo had had a generic ability to shoot and kill Franz. In his preparation for the assassination, we can imagine Gavrilo following a sequence of steps, such as (a) he purchases a gun, (b) he acquires bullets, (c) he loads the gun, (d) he draws up a plan where and when to ambush Franz, and so forth. Note that with every step taken in the period leading up to the act of assassination, Gavrilo's *generic* ability to kill transforms into his *specific* ability to actually murder Franz.

The distinction between generic and specific abilities is important, insofar as it highlights the significance of external factors, conditions, and opportunities for exercising generic human abilities. On considering this distinction, we can realise that in order to exercise or activate our generic abilities (such as the ability to read or to kill in the above examples) we often rely on external factors—factors which can be technological in nature! To be able to read we often need our glasses and something to read (perhaps, a book—which is an artefact consisting of a set of written or printed pages fastened together in one place and enclosed between softer or harder protective covers). Similarly, we normally need some kind of a weapon to be able to assassinate Austrian archdukes. On this view, technical artefacts can be attributed dispositional qualities and powers. A handgun, for example, has the capacity to fire a bullet, although it does not do so all the time, only when someone activates it by pulling its trigger. Certainly, such technical capacities are not the same as intentional human abilities which rational humans can exercise at will; that is, a handgun does not fire on 'its own volition'. Nonetheless, as bearers of powerful dispositional qualities, technological artefacts and systems do play an important role in the constitution of specific human abilities, which we shall discuss in more detail in the following section.[1]

3.2. DISPOSITIONAL POWER AND TECHNOLOGY

Having drawn a fairly detailed picture of the dispositional view of power in the preceding section, the next step should be to consider whether and how this conception of power can also be applicable to technology. While it would be highly contentious to attribute abilities—understood as intentional human dispositions—to nonhuman technological artefacts and systems, it nevertheless seems plausible that dispositional powers of some sort can be attributed to nonhuman objects, such as artefacts, machines, systems, and so on. For example, my smartphone has a feature that allows it to be used as a portable wireless hotspot. Even though this feature is always turned off and never used, the smartphone can be said to have this feature as a dispositional property. We can refer to the dispositional properties of technological arte-facts and systems as *capacities* (which is in fact a term normally used in ordinary language to refer to the dispositional qualities of artefacts). The discussion that follows aims to further elaborate on how such dispositional properties or capacities can be attributed to technology. The following passage from Quentin Gibson (1971) can serve as a good entry point:

> That a stormy sea has the power to wreck a ship or an engine the power to turn the wheels is surely something which no one should feel hesitant about main-taining. It may be said that this is anthropomorphism, and that in the literal sense only people have power. But this would be to confuse the origin of the concept with its nature. It may well be that in earlier and more animistic days, power, like causal efficacy, was attributable only to human or other spiritual agencies with wills. But generalization of the concept has taken place long since, certainly by the time of Hobbes and Locke. The move from human power to horse power, and from horse power to engine power is a matter of history. And when a physicist defines the power of an engine in terms of the amount of work it can do in unit time, he need not concern himself with its intentions. I suggest then that intentions are irrelevant. (Gibson 1971, 103–104)

This passage comes from Gibson's dispositional analysis of power, where he aimed to separate power and intention in order to develop a basic definition of power, which, devoid of intentions, would describe both human and non-human dispositional powers. As far as the passage is concerned, Gibson correctly insisted that "no one should feel hesitant" to ascribe "power" to nonhuman objects and artefacts.[2] Indeed, assuming that power is a disposi-tion of some sort (a thesis of which Gibson himself was in favour), it is possible to attribute dispositional qualities and powers to nonhuman objects and artefacts, as our earlier examples have shown. Nonetheless, to be able to attribute dispositional powers to nonhuman entities as the ones above, Gib-son is prepared to sever the connection between intentions and dispositional powers, which would then allow him to use the same dispositional concept of power for describing the powers of both human subjects and nonhuman objects.

The question we therefore should address is whether we really need the same basic concept to describe markedly different (dispositional) phenomena. Power, understood as dispositions, can refer to a number of distinct kinds of dispositions, each of which has a slightly different meaning (Morriss 2002, 27). The dispositional power of a person is radically different from the dispositional power of a thing (e.g., Taylor 1966; Ayers 1968). The same point was made earlier by Aristotle when he distinguished between *rational* and *natural* powers and dispositions (Aristotle, *Theta* 1046a–1048a). Why do we then not simply retain two different dispositional concepts (i.e., the concept of ability and that of capacity) for describing human and nonhuman powers? Gibson claims that "if it is meant . . . that there are two distinct concepts of power, this is a position which needs defending, especially as the same word is in fact used in both contexts" (1971, 101–102). Against this claim, we should note that a pluralist approach to power is not without defence (see Chapter 1). Furthermore, it is always possible to clarify whether by 'power' one means human ability or technical capacity.

Gibson's dispositional analysis of power appears to be committed to a singularist approach to power, which is manifest in his attempt to identify a single essence common to all dispositional qualities and powers of both humans and nonhumans. Such singularist position, as discussed in Chapter 1, should be resisted, insofar as it may exhibit a dogmatic assumption that there is one correct concept of power, while other concepts are incorrect. Concepts like power, as Haugaard argues, are "conceptual tools that enable us to do something. They constitute an integral part of theory, and in combination with a theoretical perspective enable us to organize information or data in a certain way, which throws the social processes involved into relief and, in so doing, deepens our understanding" (Haugaard 2012, 357). It therefore seems that a pragmatic pluralism is a better approach here, as it would support a richer conceptual and dispositional vocabulary, by retaining different dispositional concepts for describing different dispositional phenomena.

There is yet another reason why we should have (at least) two different concepts for describing human and nonhuman dispositional qualities and powers, which has to do with the attribution of moral agency and responsibility. Concepts of social power are sometimes deployed within moral and normative contexts, where they can be used for assigning moral agency or responsibility (e.g., Ball 1976; Lukes 2005). For example, a powerful social actor can be attributed forward-looking responsibility for some (morally required) future action. A powerful actor can also be attributed backward-looking responsibility for some past action (although, in this case, the reason for attributing responsibility is not one's having power, but one's having performed some action, as shown in Section 7.1). Given the moral dimension of the concept of power, using the same concept of power for both human beings and nonhuman entities can suggest an extension of moral agency to technology, which can have philosophical issues of its own (e.g., Brey 2014; see also Haugaard 2010).

The notion of technical capacity, which we said can refer to some relevant dispositional property of a technology, can however be contested by arguing that it is redundant, simply because what it aims to depict can also be described by using the concept of human ability. As we know, technical capacities and dispositions normally require human input of some kind for their activation and control. For example, the technical capacity of a car to travel from point A to point B requires a human operator who will start it at point A and stop it at point B, while controlling its movement during the time the vehicle is travelling from one point to another. The (technical) capacity of the car, one can argue, can be subsumed under, or implied by, the (human) ability of its operator to drive the car from point A to point B. Given this possibility, one can conclude, that there is no need for the concept of technical capacity. This might be a valid point, especially when we are interested in describing what the car and the driver *can do together* by forming a sociotechnical system or assemblage consisting of a human and a machine. In those contexts, where there is no need for distinguishing between what the car and the driver can do independently, a single concept of ability therefore might suffice.

Nonetheless, we can resist this argument and its conclusion on the grounds that we are living in an increasingly technological age, where we often cannot dispense with the distinction between what people can do and what technologies can do. As Hans Jonas convincingly argued in *The Imperative of Responsibility* (1984), modern technological developments have altered the nature of human action and power and thus have opened up an entire new dimension of ethical and political relevance. The anthropocentric conceptions of power and action should be thus refined and political philosophical vocabulary should be enriched with novel concepts that can better capture the changed nature of human action and power. In this regard, the concept of technical capacities is as useful a concept as that of human ability. This particularly applies to those technologies and systems whose effects and consequences cannot be easily and readily traced back either to the originators or the operators of these technologies. Consider the case of machine-learning algorithms. Traditional political and moral theory tends to view software engineers and programmers as responsible or accountable for the effects and outcomes, whether negative or positive, of algorithms. However, a primarily anthropocentric conception of responsibility would face a number of challenges, given that algorithms are becoming increasingly autonomous, adaptive, and unpredictable (e.g., Alpaydin 2010; Wallach & Allen 2010;

Floridi 2014). Therefore, concepts denoting dispositional power of technical nature can be useful after all.

3.3. TECHNOLOGY, CAPACITY AND AGENCY

After taking a closer look at some of the examples offered by Gibson in the previous section, one can notice that Gibson seems to have likened 'the power of a stormy sea to wreck a ship' with 'the power of an engine to turn the wheels' (Gibson 1971, 103). However, on some reflection, these two kinds of powers are not quite the same. Perhaps, the following episode from world history can help to appreciate the difference between the two powers. In 1274 and 1281, the Mongol ruler Kublai Khan attempted to invade and capture Japan, but during both of these military endeavours, major storms caused the sinking of a large part of the Mongolian fleet, leaving Japan safe from an imminent destructive invasion (Emanuel 2005). (At the time, to refer to these storms, the Japanese used the term 'kamikaze', often translated as 'divine wind', which, as we now know, has later been used to refer to the suicide attacks carried out by Japanese fighter pilots to destroy the Allied naval vessels during World War II). Whether the storms in question were divine or natural phenomena, no one seems to deny that their effects were of significant historical importance: on the one hand, they are seen as one of the nation-shaping events in Japanese history, while on the other hand they appear to have set a limit to the then already vast Mongol empire.

Inanimate powers of this sort could form a part of a geopolitical account of power international relations. Yet even as such many would agree that this is just a matter of luck and a happy coincidence, which was explained away as a divine intervention at the time. Of greater interest it would have been, if the Japanese had had the power to bring about storms of this scale at will—which is a feat not too far from impossible in the current era of advanced technology, considering that an underwater explosion of sufficient strength can cause a tsunami with devastating consequences (e.g., Gisler 2008). What this historical case suggests is that Gibson was wrong in likening the power of a stormy sea with that of an engine—he failed to appreciate one crucial difference between the two kinds of dispositional powers, namely that the former is untamed, its occurrence being purely a matter of natural chance, while the latter, albeit a physical disposition like the former, is controlled, its activation being mainly a matter of human will. The power of the sea and that of an engine are both dispositions, the main difference between the two is that one of them has a switch.

The preceding discussion furthermore suggests that the social and political implications frequently associated with technologies cannot stem from their technical capacities alone—there should be at the very least someone to hit the switch button. In the words of Joseph Nye Jr., "knowing the horse-power and mileage of a vehicle does not tell us whether it will get to the preferred destination" (Nye 2011, 9). The powerful implications of modern technology are often the result of the combination of technical capacities and other social and material factors, such as those pertaining to the social representations of the technology, and those pertaining to the material environment in which the technology in question is implemented. To get a better grasp of how technical capacities interact with their social and material context, we can consider different interpretations of the agency of technology. The reason why we should here consider different accounts of technological agency has to do with the fact that the philosophical and sociological literature on technology often says more about the relation between *agency* and technology rather than the relation between *power* and technology, whereby the notion of power tends to be subsumed under the category of agency. Indeed, the questions of power and agency of technology are related, insofar as the two notions are fundamentally interlinked. David Hume is one of those theorists for whom power is closely associated with agency. "The terms efficacy, agency, power, force, energy, necessity, connection and productive quality", Hume argued, "are all nearly synonymous" (Hume 1888, 157). Similarly, Anthony Giddens maintains that "power is inherently linked to agency" (1985b, 171), and that "to be an agent is to have power" (1985, 7). While Mark Warren holds that agency can be thought of in terms of "power organised as subjectivity" (1988, 59), Jiwei Ci conceives of agency as "subjectivity achieved through power" (2011, 262).

The question of whether and how agency can be attributed is no less controversial than the question of the relation between technology and power. The former has been one of the central issues in the philosophy of technology, and one that has been subject to much disagreement. According to Philip Brey (2005), there are primarily four main ways of conceiving of technological agency, such as (1) realism, (2) social constructivism, (3) hybrid constructivism, and (4) differentiated constructivism. The following subsections discuss these views in more detail and consider their implications for our understanding of technological powers and capacities.

3.2.1. Realist View of Technological Power

According to a realist conception of technological agency, the social and political effects and consequences of technologies are considered to be causally stemming from the physical and material design structure of technology. Put differently, technological agency is regarded as reducible to the physical features and properties of technology itself, and not so much to factors external to the technology (Brey 2005). In the philosophy of technology literature, there are some well-known examples which appear to endorse the realist view of technological agency and power. Consider, for example, the case of nuclear power plants discussed by Langdon Winner (1980). According to Winner, nuclear power plants necessitate strictly hierarchical and centralised management and control for their safe and appropriate implementation and operation. Winner contends that some technologies therefore are "inherently political", insofar as they can exhibit certain properties and effects in whatever setting, context, or environment. This also applies to nuclear bombs. Winner considers the nuclear bomb as inherently political since "as long as it exists at all, its lethal properties demand that it be controlled by a centralized, rigidly hierarchical chain of command closed to all influences that might make its workings unpredictable. The internal social system of the bomb must be authoritarian; there is no other way" (Winner 1980, 131). According to Brey (2005), realist accounts of technological agency and power can be problematic insofar as they tend to overlook social and institutional factors that may play a role in the realisation of their social and political agency.

3.2.2. Social Constructivist View of Technological Power

A different view of technological power can be found in what is commonly known as social constructivism (Bijker 1993; Pinch & Bijker 1987). While realism regards technological power as outcome of the physical design features and properties of technological artefacts and systems, social constructivism maintains that the purported technological agency is a consequence of how technology is interpreted and represented in society. According to Trevor Pinch and Wiebe Bijker (1987, 11–44), technology exhibits an "interpretive flexibility" which allows social actors and groups to represent or interpret the technology in different ways by attributing different properties and functions to it. Given this interpretive flexibility, different social actors and groups can ascribe different meanings and representations to a given technology. Moreover, diverging interpretations of technology can come into conflict. These instances of conflict can come to a resolution by reaching a "closure" through the process of social negotiation, whereby a specific social representation of technology becomes the dominant interpretation which then determines how the technology functions and is used. What can be problematic with the social constructivist view of technological power is that it strongly emphasises social representations and interpretations of technologies by neglecting the importance of physical properties of technologies, since there appear to be at least some affordances and constraints that derive from the physical properties of given technologies. For better or worse, some laws of physics are too inflexible for social interpretation.

3.2.3. Hybrid Constructivist View of Technological Power

A hybrid constructivist view of technological agency can be regarded as a middle ground between realist and social constructivist views discussed in the preceding subsections. Hybrid constructivism endorses the principle of generalised symmetry, according to which any elements within an analysis of social phenomena, whether they are social, natural, technical, or material, should be attributed the same explanatory role and should be analysed using symmetrical vocabulary (e.g., Law 2012, 124). The principle of generalised symmetry suggests that both human actors and nonhuman artefacts can be delegated to perform certain actions on behalf of others—that is, there is a symmetry between humans and artefacts in performing those actions (e.g., Latour 1992). In other words, hybrid constructivism rejects the traditional distinction between the material and the social, which is a distinction maintained by realism and social constructivism. Whereas realism prioritises material and physical properties of technologies in explaining their powers, social constructivism prioritises social representations of technologies in accounting for their powers. Hybrid constructivism objects to the deterministic tendencies of realism, while condemning social constructivism for treating technologies as powerless. As Callon and Latour have argued, "The choice is simple: either we alternate between two absurdities, or we redistribute actantional roles" (1992, 356). Nonetheless, the symmetrical treatment of both human and nonhuman factors in accounting for technological agency does not cohere well with more traditional conceptions of agency where human power is regarded as involving intentions which artefacts and nonhuman objects cannot possess.

3.2.4. Differentiated Constructivist View of Technological Power

It would appear that hybrid constructivism discussed in the previous subsection is the only possible alternative to realist and social constructivist accounts of technological agency. Nevertheless, as has plausibly been argued by Philip Brey (2005), there is yet another alternative account of technological power—differentiated constructivism—which avoids some of the issues faced by the other three views. For Brey (2005), the differentiated constructivism views technological agency as resulting in part from the material structure of technologies (as in realism) and in part from social processes (as in social constructivism). Moreover, differentiated constructivism holds that, although it is sometimes difficult to separate these two contributing factors (as in hybrid constructivism), such a separation of human and nonhuman factors can and should be included in giving account of technological agency. Differentiated constructivism maintains that some effects and outcomes of technological power are mainly physical in nature (i.e., they stem from the physical design of the artefact), while some others tend be social in nature (i.e., socially constructed). For this reason, differentiated constructivism finds it necessary to distinguish between the social and the technical in analyses of technology. In this manner, differentiated constructivism is an account of technological agency and power that upholds the distinction between the technical and the social, while rejecting exclusively realist and social constructivist accounts. The differentiated constructivism, as Brey (2005) argues, is not necessarily mutually exclusive in relation to the other three positions discussed above; in fact, it can sometimes play a complementary role with respect to the other three positions. Thus, with respect to realism, it highlights the role of social processes in determining the agency of technologies. With respect to social constructivism, it relates social representations and interpretations of technologies to the underlying physical processes and properties of these technologies. With respect to hybrid constructivism, it can extend and deepen the analysis by uncovering the specific distinctions that can remain hidden in a hybrid constructivist analysis. In this manner, differentiated constructivism proposed by Brey (2005), constitutes a more flexible middle ground between realism, social constructivism, and hybrid constructivism in accounting for the social agency of technological artefacts and systems.

Having so far elaborated on the four main views of technological agency, we can now apply these views to the notion of technical capacity as a dispositional view of technological power. On the realist view, technologies can be attributed technical capacities solely based on their material structures and properties. On this view, the capacity of an internal combustion engine is largely determined by the material design characteristics of the engine. But when viewed from the social, hybrid, and differentiated views of technological agency, this technical capacity of an engine alone is not sufficient in accounting for its social and political agency, in cases where such agency is attributed to it. We therefore need human abilities to properly give account for the social and political agency of technologies. Indeed, "knowing the horsepower and mileage of a vehicle does not tell us whether it will get to the preferred destination" (Nye 2011, 9).

3.4. CONCLUSION

In this chapter, we discussed the dispositional view of power and how to make sense of technological power in terms of the dispositional view of power. The dispositional view of power maintains that power is ability, capacity or potential and should be distinguished from the episodic view which defines power in terms of its exercise. When we consider technological power in dispositional terms, technology can be attributed capacities. Albeit such capacities are different from intentional human abilities, they play an increasingly important role in our understanding of the technological construction of social power. Put differently, capacities form a constitutive part of social abilities and capabilities. Indeed, our increasingly technological world requires new concepts for describing the powerful dispositional qualities of technological artefacts and systems, especially given that the effects and consequences of modern technologies cannot be easily traced back to their designers or developers.

Notes

1. Imagine for a moment what would have happened if Gavrilo's gun had actually backfired. On some counterfactual historical accounts, if Gavrilo had not succeeded in fatally wounding Frans Ferdinand, then World War I would not have happened (see e.g., Lebow 2010). Even if such counterfactual accounts are false, it is still plausible to assume that the backfiring of the gun and a resulting failed assassination attempt could have at least put off the outbreak of the war for some time. What this example suggests is that significant historical events can plausibly hinge on rather simple technical facts.

2. There has been much philosophical debate concerning the nature of dispositions as they are attributed to inanimate objects, such as the fragility of a glass, or the solubility of sugar (e.g., Carnap 1928; Mellor 1974; Mackie 1977; Prior et al. 1982; Choi & Fara 2018). While such dispositions seem to be perfectly real properties, they also appear to be "mysterious" (Choi & Fara 2014), or "ethereal" (Goodman 1954), and thus they are dissimilar to the other properties, such as those of shape and size, and thus nothing about the actual behaviour of an objects ever suggests whether it has such dispositions. To dispel this 'mysterious' appearance, there thus have been a number of attempts to analyse such dispositions. Due to reasons of space and scope, this section does not enter into this debate, since its interest in dispositions is at a slightly different level of abstraction, one that is concerned with potential differences between dispositions attributed to inanimate objects and those attributed to human actors.

Chapter Four

Systemic Power and Technology

This chapter focuses on the systemic view of power and its relation to technology. The systemic view of power defines power as a property of various social, economic, cultural, and political institutions and networks which systematically create the conditions necessary for individual or collective action. As such, the systemic view of power can usually be found in sociology, international relations theory, political economy and institutional economics, which are areas of social and political inquiry with a stronger emphasis on large-scale, transindividual social phenomena, such as societies, nations, states, markets, and so on. On the systemic view of power, powerful economic, ideological, cultural, military, and political institutions and networks are sources of social power. When viewed from the institutional lens, there can be a two-way relationship between technology and different powerful institutions. On the one hand, technology can affect existing structures of institutional power, for example, when the adoption of novel technologies reinforces or disrupts existing social or political institutions. On the other hand, existing structures of institutional power can affect the design and adoption of technology itself, for example, when a corrupt political institution removes safety requirements for the design of certain technologies. On this view, technology and different social institutions can affect one another in a mutually constitutive fashion, where changes in one can have consequences in the other.

4.1. SYSTEMIC POWER

The episodic and the dispositional views of power discussed in the previous two chapters have had some limitations. For one, the episodic and dispositional views of power tend to focus on the question of which particular agents or actors within a given polity or community have power, without giving much attention to the ways in which the structural and institutional context differentially empowers and disempowers individuals and groups within a given social system. Furthermore, the episodic and dispositional accounts of power can often be both historically and geographically restricted to the conditions found in certain contemporary democratic and affluent societies of the West. For example, the studies of the episodic conception of power, conducted by Dahl and his colleagues (e.g., Dahl 1961; Polsby 1960), primarily focused on the distribution of community power in the 1960s in New Haven, Connecticut. Similarly, the analysis of the dispositional conception of power offered by Morriss (2002) was largely concerned with the development of mathematical indices for the measurement of voting power within democratic systems of decision-making, a key feature of affluent democracies of the West. In this context, the historical account of power developed by Michael Mann over four volumes (1986, 1993, 2012, 2013) stands out in the current literature on power, not only because it offers a study of power that aims to cover the whole of human history but also because it looks beyond the individual social actors and directs our attention to the systemic and institutional sources of their power.

Mann begins his account of power by delineating his conception of human nature. He views human beings as "restless, purposive, and rational, striving to increase their enjoyment of the good things of life and capable of choosing and pursuing appropriate means for doing so" (1986, 4). Humans possessing these features are "the original source of power" (1986, 4). As such, humans, in the course of the pursuit of their goals and needs, become drawn into external relations: on the one hand, with nature, and on the other, with other humans. Thus, both "intervention in nature" and "social cooperation" are required for the satisfaction of human needs (1986, 5). Since humans are social animals – a thesis found in many thinkers, including Aristotle (1998, 4), Seneca (2011, 167), and Spinoza (2002, 338) – they achieve their mastery over nature only through social cooperation, that is, the relations of social cooperation aimed at the pursuit of power.

Such relations of social cooperation come in a variety of forms. Since there are many human goals, there are many forms of relations of social cooperation, or, as Shibutani once noted, "modern mass societies, indeed, are made up of a bewildering variety of social worlds . . . built up by people in their interaction with one another" (Shibutani 1955, 566; cited in Mann 1986, 5). Mann claims that the discipline of social theory, when faced with the challenge of studying this 'bewildering variety' of forms of social cooperation, "heroically simplifies" the task by picking out those relations of social cooperation that are most powerful and effective as a means to achieving human goals. In this manner, Mann distinguishes four main types of relations of social cooperation which people create in order to acquire power to satisfy their various needs: (1) ideological relations (involving the construction, communication, and observance of shared values and beliefs); (2) economic relations (involving the production and exchange of value, goods, and services); (3) military relations (involving the organisation of offence and defence against external forces and threats); and (4) political relations (involving the regulation of different aspects of social life through norms of justice).

These relations of social cooperation can be created on a temporary or transitory basis. However, they are frequently transformed into "institutionalized networks of interaction" (Mann 1986, 17) or "institutionalized networks of social relationships" (Giddens 1985b, 169)—meaning that such relations of social cooperation often become relatively stable networks of relations that endure in time and space. Once they become institutionalised in this fashion, they represent for Mann "institutional means of attaining human goals" (1986, 2) or, simply, institutional sources of power (or more simply, institutional power). Thus, social actors, whether as individuals or groups, derive power from the ideological, economic, military, and political institutions by partaking in these institutions. Put differently, the ideological, economic, military, and political institutions differentially empower or disempower individuals and groups. The most obvious cases of differential distribution of power can be observed in the hierarchical structure and organisation of certain kinds of institutions, such as the church, the army, or the state. For example, the church differentially empowers the pope relative to a cardinal; the army accords more power to the general than to a soldier; the monarchy entitles the monarch to more power than it does to a baron, and so on.

Mann explicitly relates his model of institutional power to the Marxist and Weberian traditions of social theory. The two theoretical traditions, Mann claims, share the common premise that "social stratification is the overall creation and distribution of power in society. It is the central structure of societies because in its dual collective and distributive aspects it is the means whereby human beings achieve their goals in society" (1986, 10). Mann also notes that Marxists and Weberians distinguish three institutional sources of power corresponding to the three core areas of social life: (1) economic, (2) political, and (3) ideological. While this three-fold classification has come to dominate contemporary social theory as the "orthodox trinity" (Poggi 2006, 135), the main disagreement lies in the interpretation of the relationship among the three sources of power. While Marxists consider the economic institution as predominant in the long run, Weberians maintain that there is no reason to think that one institution should always be more important than any of the others. In this respect, Mann is in agreement with the Weberians. However, Mann modifies the threefold model by identifying four institutional sources of power (i.e., ideological, economic, military, and political), instead of the orthodox three (i.e., economic, political, and ideological).

Contrary to Marxists and Weberians, who consider physical coercion and organised violence an integral part of the state apparatus, Mann offers four reasons for keeping the military and the political institutions conceptually separate (1986, 11). First, Mann argues that there was no monopoly of organised military power for much of human history. "Political powers are those of centralized, institutionalized, territorial regulation; military powers are of organized physical force wherever they are organized" (Mann 1986). Second, military operations are often conducted "by military groups that may be independent of their home states" (1986). Third, although it is mostly under state control, military power is institutionally separate from other state agencies, being autonomous when the military overthrow political elites in a coup d'état. One can also mention the legal autonomy of military institutions in terms of court martial jurisdiction. Fourth, under conditions of "peaceful but stratified" inter-state relations, there can be instances of "political power structuring . . . not determined by military power" (1986). This is the case, with "powerful but largely demilitarized" countries such as Japan. In this way, for Mann there are four institutional sources of power rather than the usual three.

Actually, Mann is not alone in maintaining a four-fold classification of the institutional sources of power. Anthony Giddens (1981, 1984), another eminent sociologist, defines social institutions as "the more enduring features of social life" (1984, 24), and offers a list of four major "institutional orders": (1) "symbolic order/modes of discourse" (which roughly corresponds to the institution of ideological power), (2) "political institutions", (3) "economic institutions", and (4) "law/modes of sanction" (which roughly corresponds to the institution of military power; 1984, 31; see also: Giddens 1981, 47; Mann 1986, 11). The four-fold classification of the institutional sources of power can also be found within international relations theory. Thus, for example, Russell Mead (2005, 25) identifies four kinds of systemic and institutional sources of power available to state and state actors: (1) "sharp power" (i.e., military), (2) "sticky power" (i.e., economic), (3) "sweet power" (i.e., ideological), and (4) "hegemonic power" (i.e., political) (see also Nye 2011). The subsequent section further clarifies the four forms of systemic power as discussed by Mann.

4.2. FORMS OF SYSTEMIC POWER

This section elaborates on the four major systemic and institutional sources of power as identified and developed by Mann in his four-volume account of the history of social power. These sources of power, as mentioned earlier, include the *ideological, economic, military,* and *political* forms of power. More detailed expositions of these powers are presented in the first chapters of Mann's four volumes (1986, 1993, 2012, 2013), including the second chapter of the second volume (1993). While the present section outlines the main contours of the systemic and institutional sources of power, the subsequent section will consider the role of technology vis-à-vis these four sources of power.

4.2.1. Ideological Power

Ideological power, according to Mann, stems from the human need (1) to find or give ultimate meaning in life, (2) to share norms and values, as well as (3) to take part in aesthetic and ritual practices with other humans (2013, 1–2). In this way, the institution of ideological power mainly involves the construction and communication of shared meanings, beliefs, norms and values. Ideological power is effective insofar as it assists to attain some certainty in the knowledge of the surrounding world. Ideological beliefs, as Mann notes, fill in the gaps of uncertainty with "beliefs that are not themselves scientifically testable" but that embody human fears and hopes (2013, 1). To illustrate the point, it is very difficult, if not impossible, to prove the existence of God. In this regard, ideologies can offer reassuring and supportive ideas and beliefs that make it easier for people to cope with the ever shifting social, economic, and political reality that affects them. Ideologies, Mann thus argues, "become especially necessary in crises where the old institutionalized ideologies and practices no longer seem to work and where alternatives offered have as yet no track record. That is when we are most susceptible to the power of ideologists who offer us plausible but scientifically or empirically untestable theories of the world" (Mann 2013, 1). On this view, the construction and communication of ideological beliefs and meanings come as a response or in reaction to the changes and developments that take place in the other three institutions of power (which also shows the existence of interdependence among different institutions). The institution of ideological power thus tends to be suddenly very important when humans have to face or deal with unexpected challenges and crises in the domains of human activity. This also means that ideologies can change and adapt as the problems faced by people change. Here, we should note that ideologies are not necessarily religious in character (Mann 2013, 2). While major institutions of ideology are constituted by religious beliefs, rituals, and practices, there are also ideologies that are secular in nature, such as feminism, liberalism, socialism, nationalism, racism, environmentalism, and so forth.

4.2.2. Economic Power

Economic power, according to Mann, stems from the human need to extract, transform, exchange, and consume the produce of nature (2013, 2). The institution of economic power offers a particularly powerful and effective means for attaining various human goals, because they entail, on the one hand, the "intensive" mobilisation of labour involved in the processes of production of goods, and on the other hand, the "extensive" economic relations and networks necessary for the distribution and exchange of goods or services. As such, economic relations of power can penetrate into the daily lives of most people by consuming at least one half of their waking hours, while the practices and processes of both production and exchange of goods can stretch over vast territories connecting people in the different parts of the globe. Also, economic power as a means to other ends has its own peculiar features; for example, the social change effected by economic power is seldom swift or dramatic, unlike, say, those produced by military power, and thus are slow, cumulative, and ultimately profound (Mann 2013, 2). Since the mid-eighteenth century onwards, economic power relations in many parts of the world were becoming dominated by capitalism, which treats everything involved in it—that is, the means of production, labour, and means of exchange—as commodities (Mann 2013, 2). Thus, all main elements of capitalist production and relations, including capital, labour, production, and consumption can and are traded against one another. Mann describes capitalism as "the most consistently dynamic" of forms of economic power relations in recent history, being also "responsible for most technological innovation—and most environmental degradation" (2013, 2).

4.2.3. Military Power

Military power involves the construction and exercise of power and forces necessary for the defence, offence, and deterrence against external forces. Mann defines military power as "the social organization of concentrated and lethal violence" (2013, 2). Within this definition, by 'concentrated' Mann means "mobilized and focused", while by 'violence' he means "exertion of physical force so as to injure or abuse, or intense, turbulent, or furious and often destructive action or force" (2013, 2). On this view, military power is "focused, physical, furious, and above all lethal. It kills. Military power holders say if you resist, you die. Since a lethal threat is terrifying, military power evokes distinctive psychological emotions and physiological symptoms of fear, as we confront the possibility of pain, dismemberment, or death" (Mann 2013, 2). Military power can be exercised by state armed forces, as well as paramilitaries, guerrillas, terrorists, and so on. There is frequently an overlap with state political power, but, as has been mentioned earlier (Section 4.1), for Mann, there is varying degree of autonomy between military and state political power, with military personnel normally existing as a distinct caste or class within society, subject to their own military laws, tribunals, and jurisdictions.

4.2.4. Political Power

Political power, according to Mann, involves the regulations of social life in a centralised and territorialised fashion. In other words, political power is state power, the main purpose of which is "the provision of order over a given territory" (Mann 2013, 2). In defining political power as state political power, Mann deviates not only from Max Weber, who situated political power in any organisations, but also from the notion of governance administered by diverse entities, such as corporations, non-governmental organisations (NGOs), and social movements. Instead, Mann reserves political power for state power, including the political power exercised by local, regional, and national-level governments: "States and not NGOs or corporations have the centralized-territorial form, which makes their rule authoritative over persons residing in their territories. I can resign membership of an NGO or a corporation and so flaunt its rules. I must obey the rules of the state in whose territory I reside or suffer punishment" (Mann 2013, 2). Since networks of political power are routinely regulated and coordinated in a centralised and territorialised fashion, political power, according to Mann, is more geographically bounded than the other three institutions of power. Political power can organise and regulate the other three institutions of the economy, ideology, and military. In doing so, the institution of political power can be looked at as the meta-institution that organises and regulates other institutions, as well as the relations among those institutions.

4.3. SYSTEMIC POWER AND TECHNOLOGY

Having offered an account of the institutional sources of systemic power in the preceding two sections, the next step should be to consider the role of technology vis-à-vis these forms of institutional and systemic power. Mann's historical account of the four sources of power, without doubt, has been a significant contribution to our macro-level and historical understanding of power. However, what seems to be missing is the role of technology in his account of the sources of power. To be fair, Mann certainly makes references to technology on a number of occasions. For example, in discussing the spatial contours of human societies, Mann notes that for a considerable period of the history of the civilisation, the spatial extension of economic relations in the world was largely constrained by existing transportation and communication technologies, suggesting that "only goods with a high value-to-weight ratio—true luxury items and 'self-propelled' animals and human slaves—were exchanged over long distances" (Mann 1986, 9; citing Lattimore 1962, 480–491, 542–551). However, in his historical account of power, Mann does not accord technology a more central role, with technology often being left in the periphery of history and being regarded more as a factor rather than an actor of social change. It thus seems that for Mann, power is possessed and exercised only by people motivated by their goals and needs, with technology forming the background against which social action takes place.

For a significant stretch of human history, leading up to the Industrial Revolution, the major ideological, economic, military, and political institutions may have relied less on technological means, rendering the role of technology in relation to these institutions rather negligible in the eyes of social historians and theorists. However, it is also plausible that during this stretch of history, technology may have simply remained invisible to the eyes of historians and theorists. As Lynn White Jr., an eminent scholar of medieval history, once noted, historical studies of the feudal aristocracy of medieval Europe were largely based on clerical sources that naturally mirrored the ecclesiastical attitudes, for "the knights do not speak for themselves. . . . Only later do merchants, manufacturers and technicians begin to share their thoughts with us. . . . Since, until recent centuries, technology was chiefly the concern of groups which wrote little, the role which technological development plays in human affairs has been neglected" (White 1964, V). That the significance of technology in human affairs has largely remained unmapped in history may also be due to what Andrew Feenberg has described as "the idealism of Western higher culture", which traces its roots to the aristocratic societies of the ancient Greece, where "the highest forms of activity were social, political, and theoretical rather than technical" (Feenberg 1999, 1).

For the reasons outlined above, the present section aims to consider the role of technology in the construction of ideological, economic, military, and political power. With few exceptions, the cases and examples to be discussed are mainly of historical character, which is in keeping with the historical approach of adopted by Michael Mann himself. Although the discussion that follows can be described as sketchy in comparison to the immense size of Mann's works, it nevertheless can serve as a stepping-stone for the future exploration of the topics under consideration.

4.3.1. Ideological Power and Technology

As can be seen from the preceding section, ideological power is an important form of power, at least insofar as it can offer effective means to attain goals that require subtler, relatively permanent, and lasting solutions. In this regard, ideological power resembles persuasion (see Subsection 2.2.6 and 2.3.6), as well as the concept of 'soft' power (Nye 2011). This feature of ideological power, for Manuel Castells, has mainly to do with its ability to shape human minds:

Few institutional systems can last long if they are predominantly based on sheer repression. Torturing bodies is less effective than shaping minds. If a majority of people think in ways that are contradictory to the values and norms institutionalized in the state and enshrined in the law and regulations, ultimately the system will change, although not necessarily to fulfil the hopes of the agents of social change. But change will happen. (Castells 2007, 238–239)

On this view, the construction of ideological power is the shaping of human minds, which occurs most undoubtedly through the process of communication. It would be hard to imagine ideological power without the process of communication. Yet, the process of communication is largely shaped by the particular means of communications. What could be achieved through communication one or two millennia ago is not the same as what can be achieved through communication today. Human speech, being the universal means of communication, has had at least two significant space-time limitations: in space, once the word is uttered, it can only reach those who are within the earshot; in time, once the word is uttered, "it has already vanished forever" (Schmandt-Besserat 1992, 1). It was writing, the first technological means of communication, that greatly removed these shortcomings, by making human speech relatively enduring in both time and space. By attaining spatial and temporal endurance or permanence (although relative such permanence may be), speech becomes institutionalised (which is a thesis in line with Giddens [1991], for whom endurance in time and space is a necessary condition for something to become institutionalised). This is suggestive of the role of technology in the construction and institutionalisation of shared forms beliefs and meanings.

The development of the technological means of communication did not stop with the invention of writing and the alphabet. The history of the means of communication is indeed marked by major turning points, each of which have reconfigured in considerable ways the institutions and practices of ideological power throughout the course of human history. Whether writing, printing, telegraphy, radio, television, and more recently the digital technologies, these new means of communication have offered new opportunities and constraints for the creation of shared meanings and beliefs and, therefore, for the construction of ideological power. For theoretical purposes, the history of the communication media is often split into different periods. For instance, Terence Moran (2010) writes about the different turning points in the history of the development of communication media. According to Moran, with the invention of writing, humans became *literate*; with the advent of printing, humans became *typographic*; with the arrival of graphics, such as photography and cinematography, humans became *hypergraphic*; with the development of electrographic and electrophonic communications, humans became *electric*; and with the advance of digital communications, humans became *cybernetic*. Similarly, Marshall Poe (2011) distinguishes five different periods of the social history of communication: (1) *homo loquens* in the age of speech; (2) *homo scriptor* in the age of manuscripts; (3) *homo lector* in the age of the print; (4) *homo videns* in the age of audio-visual media; (5) *homo somnians* in the age of the internet.

Let us consider the case of the radio and television broadcasting, which, having emerged in the early years of the twentieth century, still remains as an important tool for the shaping of the public mind today. The advent of the radio and television as a means of communication gave rise to the so-called mass communication, since they made it possible to broadcast information to larger and larger audiences. Castells describes radio and television media as "vertical" means of communication (2007, 247). Similarly, Bernstein characterises them as "the most hierarchical" of all the communications technologies, since "no preceding media could reach so many people so instantaneously and with so little feedback in the opposite direction" (Bernstein 2013, 220). The capacity to broadcast information to masses proved a very effective way of shaping human minds, and an effective tool for propaganda. Bernstein documents that during the period between 1920 and 1980—the years dominated by the radio and television media—the world witnessed a sharp rise in the number of totalitarian regimes and states. Certainly, this correlation by no means implies causation, since there were other causal factors at work, such as the disastrous consequences of World War I, as well as the subsequent global economic depression. Nevertheless, while propaganda is one of the most powerful instruments of totalitarianism (Arendt 1973), the arrival of hierarchical and vertical means of communication well suited for the purposes of propaganda, can certainly count as one of the major factors in explaining the spread of totalitarianism in the last century.

As is known, television (and to a lesser extent radio) broadcasting has survived even in the digital age of the internet. Its continued existence appears to have been facilitated by the advent and adoption of cable (Smith 1970), digital (Layer 2001; Galperin 2004), satellite (Puente 2010), and internet Protocol Television (IPTV) (Jukes et al. 2018) methods of transmission of televisual content, which have also greatly improved the picture and sound quality of television broadcasting. Nowadays radio- and television-enabled mass communication remains one of the main channels of communication between the political system and the population (e.g., Castells 2007, 2009). Even in the United States, a country with one of the highest rates of internet use, television remains one of the most frequently used sources of news (Bennett 2005; Ferguson & Greer 2016; Robinson et al. 2018; Jukes et al. 2018).

In 1988, Edward S. Herman and Noam Chomsky published their landmark book *Manufacturing Consent* which inquired in the role of mass media practices in the construction of ideological power. As the word 'manufacturing' in title of the work suggests, throughout much of the twentieth century the mass media was founded on a model of mass production that resembled the conveyor belt assembly lines and railways of the industrial society, in which a centralised authority oversaw the production and distribution of messages from one major hub to the rest of the region, country, or world. In *Manufacturing Consent*, Herman and Chomsky aim to explain, with well-documented case studies, how propaganda enters the mass media and thus shapes public opinion in democratic societies, in the absence of covert or overt practices of censorship and coercion present in authoritarian regimes, where anyone who fails to deliver the official ideology or dogma simply gets imprisoned, beaten, raped, or killed. According to these authors, the mass media were defined by the certain specific communication technologies, such as the newspaper, radio, and television, all of which were capable of mass producing and broadcasting texts, sounds, and images for mass public consumption. For this reason, the authors develop a "propaganda model" that consists of a set of five news "filters" that can screen out 'unacceptable' ideas without the use of coercion and force (Herman & Chomsky 1988, 2):

(1) "The size, concentrated ownership, owner wealth, and profit orientation of the dominant mass-media firms". The costs involved in running even very small newspapers has always exceeded hundreds of thousands of dollars. Hence "the first filter—the limitation on ownership of media with any substantial outreach by the requisite large size of investment". (1988, 3–14)

(2) "Advertising as the primary income source of the mass media". "The advertisers' choices influence media prosperity and survival. The ad-based media receive an advertising subsidy that gives them a price-marketing-quality edge, which allows them to encroach on and further weaken their ad-free (or disadvantaged) rivals". (1988, 14–18)

(3) "The reliance of the media on information provided by government, business, and 'experts' funded and approved by these primary sources and agents of power". The mass media are typically drawn into a "symbiotic relationship" with powerful sources of information through "economic necessity and reciprocity of interest". (1988, 18–26)

(4) "'Flak' as a means of disciplining the media". Here flak is defined as "negative responses to a media statement or program. It may take the form of letters, telegraphs, phone calls, petitions, lawsuits, speeches and bills before Congress, and other modes of complaint, threat and punitive action". (1988, 26–29)

(5) "'Anticommunism' as a national religion and control mechanism". Those "accused of being pro-Communist or insufficiently anti-Communist, are kept continuously on the defensive in a cultural milieu in which anticommunism is the dominant religion" (1988, 29–31). (In the post–Cold War era, anti-communist sentiments may have been taken over by pro-American sentiments).

The five filters, according to the authors, interact with and reinforce one another. "The raw material news must pass though successive filters, leaving only the cleansed residue fit to print" (Herman & Chomsky 1988, 2). In this manner, the five filters narrow the range of news that passes through their screens, sharply limiting what can be presented as 'big news'. News from political and business establishment sources meets one major filter requirement and is willingly accommodated by the mass media. In contrast, messages from and about the marginalised individuals and groups, whether domestic or foreign, are put at a disadvantage vis-à-vis costs, and credibility, while being in potential disagreement with the interests of the gatekeepers and other powerful parties. What Herman and Chomsky have described as filtering is similar to what Bennett (1990) has termed as "indexing", which is the practice of limiting the range of political issues and views to be reported to those articulated within the mainstream political establishment.

The practices of filtering and indexing the news are reminiscent of the second face of power as theorised by Bachrach and Baratz (1970; see also Section 2.1). In their critique of the first face of power, Bachrach and Baratz argued that Dahl and his colleagues looked only at the surface of democratic politics, and that there is a different dimension of power, its second face, that pertains to the manner in which some political issues are organised into politics, while other are organised out. According to Bachrach and Baratz, certain important issues do not reach the surface, because the political elites make sure that these issues do not get to be mentioned. The account of the distribution of power in New Haven, Connecticut, offered by Dahl relied on the examination of those political issues that were discussed in the local press. Since neither poverty nor race relations were mentioned in the press, these issues were left unexamined by Dahl. Issues relating to race relations however turned to be important ones, since, around the time of the publication of Dahl's work, race riots took place across many U.S. cities, including New Haven. While the political establishment is careful about keeping certain critical issues off the agenda, the "filtering" and "indexing" practices of the media results in limiting the range of issues to those articulated within the political establishment. As a result, what is not reported in the media and the press tends not to reach the public mind.

The filtering practices identified by Herman and Chomsky, and the indexing methods discussed by Bennett may not easily apply to the new realities of the internet-based media. Nonetheless, television-enabled broadcast news remains a powerful instrument of political communication. In *Power of Communication* (2009), Castells notes that a contemplation of the process through which the United States began the war in Iraq would show that mass media-based state propaganda is still a powerful force in shaping public opinion, to a greater extent because the traditional broadcast media remain the dominant media today. Television still remains the central medium through which people get their information about the world. It is indeed frequently the case that the news items discussed on the blogosphere are the things people have read in newspapers, heard on the radio, or seen on television (Castells 2009).

4.3.2. Economic Power and Technology

One of the features that technology and economic power have in common is that mainstream economics has long overlooked them (e.g., Hessler 2014). While economic power involves using economic means (such as wealth, resources, etc.) to bring about desired political outcomes, the role of technology in creating this kind of economic power can be described as a two-step sequential process: first, technology plays a major role in bringing about economic progress by driving growth and creating wealth; and second, the resulting economic resources and wealth become an essential means of achieving political goals (e.g., Dickson 2008).

From the start of the first Industrial Revolution, technology has been a major driving force of economic growth. However, it is only recently that technology has rightfully been recognised as an important engine of economic growth (e.g., Solow 1965, 1970; Mokyr 1990; Grossman & Helpman 1991; Aghion & Howitt 2009). Beginning from around the time of the Industrial Revolution, a network or constellation of societies that has come to constitute the so-called 'Western Civilisation' emerged within a position of economic and technological superiority over that of the rest of the world. This has been somewhat remarkable, given the earlier advantages of the economies of Near and Eastern Asia. To explain this shift, economic historians have proposed different theories, including what might be called 'exceptionalist' theories of European science, population mentality, and socio-political organisation, while others have attempted to give account of this shift by appealing to some such factors as geographic determinism, natural resource distribution, and so forth.

Against this backdrop, Joel Mokyr, in his landmark work, *The Lever of Riches* (1990), has argued that the previous emphasis placed by economic historians on the 'Smithian growth model' (which maintains that economic growth is achieved through expanded commerce) (Mokyr 1990, 5), and on the 'Solovian growth model' (according to which economic growth is a result of more capital investment) (Mokyr 1990, 4) downplays an important factor in the economic ascendancy of the Western Civilisation. Neither kind of explanations, Mokyr has claimed, can account for the almost explosive economic growth of the West between the fifteenth and the nineteenth centuries. Rather, Mokyr argues, the relative surge of the Western economies should more properly be seen as a result of the 'Schumpeterian growth model', according to which economic growth and development is regarded to be a product of technological change. According to Mokyr, the "stock of human knowledge" led to the development and expansion of technologies brought about the advances in technology that in turn allowed efficient production of more and better goods as well as efficient utilisation of existing resources (1990, 6). Thus, in developing his explanation of the economic ascendancy of the Western world, Mokyr has focused on the means of growth, by examining the process of technological change from the period of the Ancient Rome through to the dawn of the twentieth century, as well as inquiring into the reasons why certain societies or regions exhibited "technological creativity", while others did not (1990, 11; see also Section 9.2).

According to Mokyr, there have been two explanatory models of technological change in economic historiography. On the one hand, there have been "heroic theories of invention", which explain technological change as resulting from the activity of handful "individual geniuses" (Mokyr 1990, 12); while on other hand, there has been the "drift theory", which aim to explain technological change on the basis of "anonymous, small, incremental" technological improvements and refinements (Ibid.). Having distinguished between the two models, Mokyr contends that both models fail to wholly explain technological change of the past. On the basis of historical evidence pertaining to the past two millennia, Mokyr argues that a proper model of technological change should combine both approaches. Although much of technological change was gradual and incremental in much of Western European history, it is hard to ignore a few revolutionary and ground-breaking inventions. To these pioneering inventions, Mokyr includes, for example, the "profoundly brilliant" invention of the printing press by Johannes Gutenberg (1990, 49), and the invention of technology for producing cheap steel by Henry Bessemer (1990, 116; see also Gale 1973). These kinds of ground-breaking inventions Mokyr calls "macroinventions" (1990, 291), while continual technological modifications, Mokyr terms "microinventions" (1990, 292). In this manner, Mokyr argues that technological progress takes place when societies can maintain both kinds of inventions: in the absence of macroinventions technological progress slows down and comes to a halt, whereas without microinventions great ideas and innovations may not fully manifest their ground-breaking potential.

Mokyr offers a rather strong, empirically well-supported, case, for the thesis that both macroinventions and microinventions played a crucial role in bringing about the economic dominance of the West. Mokyr has presented a rather strong argument that technological creativity–driven technological change was, if not a cause, but rather a causal condition, or a contributory factor in the economic rise of the West. As can be read, Mokyr refrains not from admitting the complexity of the picture, while at the same time acknowledging the research that still remains to be done.

There have in fact been a number of prominent empirical studies of the role of technological innovation and industrialisation in driving economic growth in different countries. One subset of these studies pertains to a set of countries which emerged as economically backward countries in the after-math of the World War II, and which rose economically through industrial-isation (Amsden 1992, 2001). Often referred to as 'late' or 'newly' industri-alising countries, these include Japan, South Korea, Taiwan, Malaysia, Bra-zil, Turkey, India, and Mexico (Amsden 2001). Although these counties have certain substantial differences, they share one feature in common—they be-gan to industrialise their economies on the basis of borrowing, and learning from, foreign technology in order to catch up with the more industrialised countries within the global market (Amsden 1992).

Formulaically, the role of technology as an engine of economic growth can be described in the following manner (de La Grandville 2008, 487). At the beginning of year Y, society S has in its possession a capital reserve Cy and labour force Ly. During the year Y, on the basis of its capital and labour resources Cy and Ly, the society creates a product, which is referred to as Gross Domestic Product (GDP). Part of this GDP is then utilised to replace the capital reserve Cy which has depreciated during year Y. After subtracting the replacement costs, society S ends up with Net Domestic Product (NDP). This NDP is then divided into two parts: (1) consumption (e.g., 70% of the NDP), and (2) investment (e.g., 30% of the NDP). This investment will be used to increase the capital reserve Cy+1 at the beginning of the year Y+1. As a matter of course, during the year Y, the labour force can grow from Ly to Ly+1, and technological capabilities can rise to new levels. These in-creases in labour force and technological capabilities will then enable the society in question to produce a higher GDP, and therefore a higher NDP, in the year Y+1. As can be seen, each future annual increase in income will hinge on the amount of investment made, as well as the increases in the labour force and technological capabilities made in each preceding year (de La Grandville 2008, 487). Labour=saving technology is an important engine of economic cycles and economic growth.

While the aforementioned studies consider the role of technology as an engine of economic growth in the process of industrialisation of economies, this role of technology has also become more prominent in the process of informatisation of economies in the last five decades or so. Roughly begin-ning in the 1970s, there has been a shift from industrial society to postindus-trial information society, where the production and utilisation of information occupies a dominant position within the social and economic realms (e.g., Bell 1973; Beniger 1986; Harvey 1989; Webster 1995; Castells 2010a). For example, according to Bell (1973), the postindustrial society is the stage of social development where information becomes a valued form of capital, production of knowledge becomes one of the main drivers of economic growth, and the role of professionals in creative industry and information technology domains rise in value. Beniger (1986), who has also studied the move from industrial to information society, has contended that the information revolution of 1970s has come about as a solution to the crises of control brought about by the constraints and limitations of the means of transportation and communication in an increasingly individualising and diversifying world. Similarly, Harvey (1989) has argued that the postmodern economic system has brought about a culture of speed, which makes the processes of production and consumption very dynamic by embracing commodification and heterogeneity. Likewise, Webster (1995) maintains that the new communication technologies have made possible the transitions from Fordist to post-Fordist way of production that is more efficient and adaptable with regard to its globalising environment. Castells (1996) also holds that the shift from industrial to information society is largely a result of the transformation of the capitalist economic order through information technology. While in the industrial society the basic unit of economic organisation was the factory, in the new information society this unit is the network, consisting of individuals and organisations and continually adapting itself to its market environment.

The major technological core of this new information society is undoubtedly the internet, which is now seen as one of the driving forces behind contemporary economic globalisation (e.g., Van Dijk 2006; Lessig 2001; Fuchs 2008). Thus, under current conditions of informatisation and globalisation of economic relations, the so-called global technology giants, such as Apple, Google, and Microsoft, have emerged as the wealthiest and economically powerful non-state entities (e.g., Morozov 2015, 2016). Indeed, as noted in a recent report by the ratings agency Moody's, the amount of cash held by the U.S. non-financial companies reached $1.68 trillion (USD) at the end of 2015. The top five companies that had the most cash were IT giants Apple, Microsoft, Google, Cisco, and Oracle. These technology giants together possessed $504 billion (USD), while Apple alone held $216 billion (USD) (Treanor 2016). Thus, in the contemporary information society, where information technology is a vital engine of economic growth and globalisation, IT giants can amass vast amounts of economic wealth, and, for better or worse, wealth can be an instrument of achieving political objectives.

4.3.3. Military Power and Technology

Colin Gray, formerly a member of the Reagan administration, currently a professor of international politics at the University of Reading (United Kingdom), writes that "the political history of the world may be told in terms of changes in military technology" (Gray 1975, 238). A classic illustration of this thesis can be found in Lynn White's *Medieval Technology and Social Change* (1964) which was a study of the social consequences of the introduction and widespread adoption of the stirrup in the medieval period. Although this particular thesis may not be true of the political history of the world as a whole, it seems nevertheless applicable to the history of warfare, especially when it comes to its theoretical periodisation.

The importance of technology in the construction and development of military power can indeed be observed in the way in which historians have often interpreted the history of warfare in terms of different periods associated with the advent of certain kinds of technologies. For example, history of warfare, more traditionally, has been divided into periods associated with the use of certain specific kinds of weapons, such as 'the age of edged weapons', 'the age of gunpowder', or 'the age of the nuclear bomb', and so on. The destructive and lethal aspects of military power have indeed been subject to continuous technological enhancement and sophistication; or in the words of Saint Augustine: "for the injury of Man how many kinds of poison, how many weapons and machines of destruction have been invented" (Saint Augustine 2009, 768). Although such historical periodisation is suggestive of the role that technology plays vis-à-vis military power, they nevertheless tend to restrict this role to lethal technologies and weaponry only. Since effective military power requires more than just the means of killing and destruction, one must not underestimate the importance of, for example, transportation and communication technologies in the achievement of certain specific military objectives. Effective military power requires both lethal and non-lethal technological capacities. For example, in 2014, in his plea to the U.S. Congress, the president of Ukraine urged the U.S. lawmakers to assist his country by providing "military equipment, both lethal and non-lethal. . . . Blankets and night-vision goggles are [also] important" (quoted in Payne 2014).

For this reason, Martin van Creveld, a noted military historian, has proposed a periodisation of military history that can "reflect the development of technology as a whole" (van Creveld 1991, 2). In his *Technology and War* (1991), van Creveld divides the history of warfare into four primary technological periods. The first period, which covers the years between 2000 BC and 1500 AD, van Creveld calls "the age of the tool", since this period is characterised by the prominence of technologies the operation of which required the muscle energy, whether human or animal. From Archimedes's pulleys to Leonardo's crank-driven war machines, these tools rely on muscle power. Some of the most prominent military inventions of this period, too, exhibit the same logic: the spear (e.g., Dohrenwend 2012), the bow and arrow (e.g., Bradbury 1985), the sword (e.g., Bradford 1974), the galley (e.g., Morrison 1995), the chariot (e.g., Littauer 1972), the stirrup (e.g., White 1964)—are all technologies the operation of which depended on either human or animal power.

The second period, which begins in the Renaissance and ends slightly before the second Industrial Revolution, van Creveld calls "the age of the machine" (1991, 81), given that in this period technology had reached the point where a prominent role was being played by machines which, unlike those from the age of tools, did not require the human or animal muscle energy, but the energy derived from the wind, water, and the explosive force of gunpowder. Some of the important weapons of this period were the emergence of the early guns and canons. As van Creveld notes, a gun or a canon of whatever size had the same underlying logic of "an internal combustion engine acting in one direction instead of two" (1991, 82). And it is in this respect that guns and canons are machines rather than tools. The development of such gunpowder-based weapons thus created the technical capability to hurl heavier projectiles with much greater force than the largest catapults or trebuchets of the tool age.

The third period, which starts slightly before the second Industrial Revolution and ends with World War II, van Creveld calls "the age of the system" (1991, 153–232). This is the period in which isolated machines and artefacts came to be integrated into complex technological systems "whose parts depended for their functioning—their existence, even—on precise coordination with the rest" (1991, 3). The railway, the telegraph, and the telephone are some of the most prominent technologies of this period. Within the field of technology studies such systems are generally called socio-technical systems, and/or large technical systems (e.g., Joerges 1988, 9–36; Hughes 1983, 51–82).

Finally, the fourth period, which begins with the end of World War II and continues up to the present day, van Creveld has termed "the age of automation" (1991, 235–310). Although this period witnessed the use of the first nuclear weapons, van Creveld argues that "in retrospect the advent of cybernetics and feedback seems to be even more important" (1991, 4) than nuclear weapons. The age of automation is a historical period in which warfare is conducted with the help of machines that are "capable, within limits, of themselves detecting changes in their environment and of reacting to those changes" (1991, 4).

Certainly, the historical periodisation offered by van Creveld is to some extent arbitrary, insofar as the inventions and technologies that emerged in one period were still being used in the other periods. Put differently, there has been considerable overlap between these periods. Thus, for example, technologies relying on the non-human or non-animal power, such as a sailing ship that relied on the power of the wind, were in use during the age of the tool, long before the age of the machine. Or, some of the technologies, having originated in the earlier periods continued to be used in the later periods. Yet, it might still be said that what justifies the naming of these periods is not whether or not a certain kind of technology is being used but their prominence, as well as their breaking down the moulds of the past.

4.3.4. Political Power and Technology

Since the notion of political power, within Mann's framework, is reserved for state political power, to get an understanding of the role of technology in the construction of political power we should consider the role of technology in the emergence of the modern state. According to Joseph Strayer (1973), the emergence of the modern state began in Western Europe in the period between 1100 and 1600 AD (1973, 10). This by no means suggests that there were no states outside of Europe, or in the earlier periods—certainly the Persian, the Roman, the Han Empires, as well as the Greek polis were all states. What Strayer means is that the modern state did not originate directly from these earlier examples, because, firstly, the founders of the early modern states knew very little of the earlier states, since they were far removed from them both in time and space; and, secondly, the type of state that Europeans had to reinvent through their own efforts, eventually proved to be more successful than the earlier models of states (1973, 11). The vast empires were largely tied together through military power and therefore did not enjoy the basic loyalty of their inhabitants, especially in the far-flung territories of the empire (1973, 11). The city-state or the polis, although they enjoyed the basic loyalty of their subjects, were militarily weak and therefore frequently became victims of conquest (1973, 11). In contrast, the early modern states of Europe, Strayer argued, managed to combine the strengths of both the empires and city-states—they both ensured the basic loyalty of their citizens and were sufficiently large and powerful to have excellent chances for survival (1973, 12).

The difference between modern and premodern states, according to Strayer, is further suggested by the emergence of political institutions with the following four features:

(1) They persist and endure in time and space. According to Strayer, in order to become a state, a human community must persist in time and space. Temporary relations of cooperation cannot become the basis for a state, unless the causal conditions that give rise to these relations of cooperation last sufficiently long or recur frequently enough that the relations of cooperation become permanent. Moreover, there should be a certain geographic area within which the social community can found its political system. (Strayer 1973, 5–6)

(2) They are impersonal. Temporary and primitive political groupings can function through personal relationships, such as the meetings of the heads of families. However, in order for a political community to persist in time and space, there ought to be impersonal political institutions which "can survive changes in leadership" and "allow a certain degree of specialisation" (1973, 6–7). Thus, in modern states, authority rests not with the person who exercises it, but with the role or position that this person occupies (see also Subsection 2.2.5).

(3) They are the source of final authority. According to Strayer, in addition to being impersonal and permanent, political institutions should come to possess final authority that can give final decisions that cannot be reversed by any other authority in a given territory (1973, 7–9). It can further be added, that to become final authority, judicial political institutions would require a certain degree of centralisation (Strayer 1973, 14). This need for centralisation is also in accord with Mann's conception political power as discussed in Subsection 4.2.4.

(4) They command the basic loyalty of the population. According to Strayer, while in premodern states, people's loyalties rested in their family, local community, or religious organisation, the modern states required the shift of such loyalties to the state. This meant that the state came to require a form of ideology or moral authority to justify its institutional structure, in such a way that subjects should accept the view that "the interests of the state must prevail", and that "the preservation of the state is the highest social good". (1973, 9–10)

The four requirements specified by Strayer are interconnected and overlap to a certain degree. For example, in order to command the basic loyalty of its subjects or citizens (requirement 4), the state must ensure that there are no other competing or conflicting authorities within its territory and that only the state is the final authority (requirement 3). In order to become the final authority within its territory (requirement 3), political institutions need to become centralised and, more importantly, impersonal (requirement 2), so as to make it possible for the state, for example, to remove noncompliant officials from their posts. Finally, in order for the political institutions to become impersonal—to be able to survive changes in leadership (requirement 2)—the state political institutions must persist or endure in time and space (requirement 1); and vice versa: state political institutions begin to obtain endurance (requirement 1) by becoming impersonal (requirement 2).

We can now consider the role of technology in the emergence and development of modern state institutions characterised by the four requirements identified by Strayer. With regard to the first requirement (i.e., institutions must persist in time and space), we can note that technology can grant endurance to institutions, making them more durable in time and space. In 'Do Artifacts Have Politics?' (1980), for example, Winner has argued that technological innovations "establish a framework for public order that will endure over many generations" (Winner 1980, 128). Similarly, Latour has argued that to in order to better understand issues of power and domination, it is necessary to "weave" social and political relations "into a fabric that includes non-human actants, actants that offer the possibility of holding society together as a durable whole"; hence "technology is society made durable" (Latour 1991, 103).

With regard to the second requirement identified by Strayer (i.e., political institutions must be impersonal), it can be argued that technology can help political institutions to achieve a certain degree of being impersonal. Generally speaking, technology can help make social relations impersonal in two different ways: on the one hand, by replacing human actors with technical artefacts, devices, and machines, as in Latour's well-known example of non-human grooms in his essay 'Where Are the Missing Masses?' (1992), and on the other hand, within socio-technical systems, by replacing a human actor with another human actor, just like the replaceable or interchangeable parts of large machinery (this latter phenomenon is somewhat similar to the notion of alienation [e.g., Seeman 1959]).

With regard to the third requirement (i.e., political institutions must become the final authority), it has already been noted that the emergence of such final authority within a given geographical territory requires a certain degree of centralisation of political institutions within that territory. The inalienable part of any attempt to centralise political agencies and organisations is, without doubt, the establishment of communication and transportation infrastructure. The construction of transportation and communication lines has always taken centre stage in building and sustaining a strong state apparatus, even outside medieval Europe. For instance, in nineteenth-century Iran, Nasir al-Din Shah, the ruler of Iran, installed the first telegraph lines throughout his country in order to facilitate the process of receiving information even from the low-ranking officials in the remotest villages (Rubin 2001, 19). This eagerness for the adoption of the telegraph, as Morozov (2011, 91) correctly notes, echoes the advice given by Iran's eleventh-century minister Nizam Al-Mulk in his renowned *The Book of Government or Rules for Kings* (2002), according to which rulers and kings should have dual lines of communication and information.

Finally, when it comes to the fourth requirement (i.e., political institutions must command the basic loyalty of the population), it can be argued that technology, in particular different technological means of communication, has played a crucial role in spreading state ideologies that legitimise their political structures and institutions. It is indeed for this reason that means of communication has frequently been the object of state control or monopoly. Besides communication technologies, the state could rely on technological means of violence and coercion in maintaining its legitimating ideology. Such means of violence and coercion would come in handy when, in seeking the loyalty of its population, the state would frequently need to curb the influence of competing authorities within its territory, such as the local lords, powerful families and clans, religious groups, or communities.

4.4. CONCLUSION

In this chapter, we have explored the systemic aspect of power and its relations to technology by analysing discrete institutions including the ideology, the economy, the military, and the polity or the state. Individuals and groups partaking in these institutions are differentially empowered and disempowered, while they lose certain powers (e.g., as a result of institutional constraints), they also gain certain new powers (e.g., as a result of institutional affordances). By conceiving of power as a property of various social, economic, cultural, and political institutions and networks which systematically create the conditions necessary for individual or collective action, the systemic view of power is usually found in sociology, international relations theory, political economy, and institutional economics, which are areas of social and political inquiry with a stronger emphasis on large-scale, transindividual social phenomena, such as societies, nations, states, markets, and the like. When viewed from the institutional lens, there is a bi-directional (or, a two-way) relationship between technology and institutional power: (1) in one direction, we can theorise how technology can affect existing structures of institutional power, for example, when the adoption of novel technologies reinforces or disrupts existing social or political institutions; (2) in the other direction, we can theorise how existing structures of institutional power affect the design and adoption of technology itself, for example, when a corrupt political institution removes safety requirements for the design of certain technologies. On this view, technology and different social institutions can affect one another in a mutually constitutive fashion, where changes in one can have consequences in the other. The main problem with this way of theorising is that by focusing on discrete institutions we may lose sight of discreet power, that is, the more invisible or insidious ways in which power can come to shape and constitute the very selves and identities of individuals. This problematic will be the focus of the following chapter that discusses this constitutive character of power.

Chapter Five

Constitutive Power and Technology

This chapter concerns the constitutive conception of power and its relation to technology. The constitutive view of power conceives of power as not only structurally creating the conditions necessary for individual or collective action but also as systematically constituting the social actors themselves. The constitutive view of power has been particularly prominent within poststructuralist social and political theory, which considers, among other things, how our language can set limits to what it is possible for us to think, thereby shaping who we are and how we behave. The present chapter examines some of the main poststructuralist implications for our understanding of power and technology, by considering Michel Foucault's contributions to this intellectual domain. Foucault is widely regarded as one of the most renowned and influential poststructuralist thinkers of the last century. His studies of power mark a significant departure from the views of power considered in the preceding chapters. Power, according to Foucault, is not so much what is possessed or exercised by individuals but also that which is productive and constitutive of selves, identities, characters, and persons. The discussion further highlights how technology has become implicated in processes of production and constitution of selves and identities.

5.1. POSTSTRUCTURALISM AND POWER

In September 2014, Apple Inc. launched its Health Kit application that was designed to allow iPhone users to track a very wide range of bodily metrics, including height, weight, temperature, levels of alcohol intake, and so on (Lewis 2014). As stated by Craig Federighi, a senior Apple Software executive, "with Health, you can monitor all of your metrics that you're most interested in" (quoted in Duhaime-Ross 2014). However, to much surprise, the application did not offer women the opportunity to monitor their menstruation cycles. In fact, it took a significant amount of time for the company to address this design omission, even after it had been pointed out by many commentators (Newman 2015).

The question that needs addressing here is whether this design omission involved an exercise of power, say, by designers over users. While we may never know the precise reason for the design omission, let us assume for the sake of an argument that the design omission was not deliberately intended on the part of the designers (it is quite plausible to imagine designers excusing their neglectful design by claiming that it was not intentional). For those theorists, who restrict the meaning of power to intentional actions (e.g., Dahl & Lindblom 1953; Wrong 1968; Morriss 2002), the above case of design omission would not constitute an exercise of power, because the designers did not intend to omit the feature in question. However, this restrictive attitude to power relations would be regarded as rather ingenuous by (at least) most feminist thinkers, who would argue that the social context in which technological design takes place is itself subject to patriarchal structuring of power relations, where the interests of women are subordinated to the interests of men. Such patriarchal structuring of power relations, according to Chris Weedon, a feminist poststructuralist, is an implication of patriarchal discourses where different meanings and values are attributed to our biological sexual difference:

> In patriarchal discourse the nature and social role of women are defined in relation to a norm which is male. This finds its clearest expression in the generic use of the terms 'man' and 'he' to encompass all of humankind. . . . To say that patriarchal relations are structural is to suggest that they exist in the institutions and social practices of our society and cannot be explained by the intentions, good or bad, of individual women or men. (Weedon 1987, 2–3)

On this understanding of power relations, the above omission of "one of the most basic metrics of human existence" (Duhaime-Ross 2014) would thus constitute a form of power relationship, without any regard to the good or bad intentions of the designers in question, because such particular intentions themselves would be interpreted as the product of patriarchal discourses that have significant implications for power in society. In fact, an often-cited reason for a gender-biased design omission is "a corporate culture dominated by men" (e.g., Michelfelder et al. 2017, 193), where designers, who are often men, as a matter of habit think of other men as target users for their products. As such, the often-cited 'corporate culture dominated by men' is a particular instantiation of the patriarchal structuring of power relations in society. An appropriate analysis of such power relations, according to Weedon, requires a poststructuralist theory "which can explain how and why people oppress each other, a theory of subjectivity, of conscious and unconscious thoughts and emotions, which can account for the relationship between the individual and the social" (Weedon 1987, 3). Poststructuralist theory, as Weedon envisions, would go beyond the actual or putative intentions and experiences of participants of power relations, and would aim, instead, for an explanation of the very power relations involved in the formation of these subjective intentions and experiences. Poststructuralism refers to

> a theory, or a group of theories, concerning the relationship between human beings, the world, and the practice of making and reproducing meanings. On the one hand, poststructuralists affirm, consciousness is not the origin of the language we speak and the images we recognize, so much as the product of the meanings we learn and reproduce. On the other hand, communication changes all the time, with or without intervention from us, and we can choose to intervene with a view to altering the meanings—which is to say the norms and values—our culture takes for granted. The question is just the one . . . who is to be in control?" (Belsey 2002, 5)

On this perspective, the very act of restricting the definition of power to cases of direct and intentional exercise of power itself is regarded as an exercise of power. By limiting the senses of the term 'power', we can, knowingly or unknowingly, consign to oblivion those effects of power which are not covered by the restricted definition of power. As Wrong himself acknowledged: "efforts to influence others often produce unintended . . . effects on their behavior. . . . The effects others have on us, unintended by and even unknown to them, may influence us more profoundly and permanently than direct efforts to control our sentiments and behavior" (Wrong 1968, 676). Dahl and Lindblom refer to such unintended effects of power with a rather technical phrase: "spontaneous field control" (1953, 99; which is a rather technical phrase befitting more obscure psychology textbooks, one could say). Nonetheless, a simple search on Google Scholar can show that there are definitely more academic works with 'theory of power' in their title than those titled 'theory of spontaneous field control'. Without making recourse to singularism in defining power (which, as noted in Chapter 1, can lead to dogmatism, relativism, or scepticism about power), we can describe power as that which can affect individuals in different ways, in unintended ways in the process of their socialisation, and in intended ways within interactions of direct social control. In this respect, one of the benefits of poststructuralism is that it allows us to focus on the practical, moral, and political implications of different ways of theorising about power.

Our task in the following sections is to explore some of the main implications of poststructuralism to our understanding of power and its relation to technology, by considering the theoretical contribution made by Michel Foucault to this strand of thought. Foucault is by far one of the most influential poststructuralist thinkers of the past century, whose influence on our understanding of power reverberates to this day. We should note that approaching Foucault on power might present some challenges. First, as Wickham (1983) notes, Foucault's thought on power is spread across numerous fragmented sources, including articles, lectures, and interviews. Second, as Jon Simons notes, "commentary on and critique of Foucault's notion of power has become an intellectual industry in itself" (1995, 129; see also O'Farrell 1982, 449). Finally, some have found issues in Foucault's use of (French) language (e.g., Steiner 1971; Hacking 1981; Taylor 1984, 90), which can be further exacerbated by its translation into English (Aron 1964, 255). However, despite the issues of translation and interpretation, Foucault can be regarded as an author whose work can be placed "in a line of scholarship in which Weber would not be unrecognizable" (Clegg 1989, 155).

5.2. CRITIQUE OF SOVEREIGN POWER

Michel Foucault was a French philosopher who wrote extensively within the areas of social history and philosophy. Although the works of Foucault consider apparently diverse social and historical subject matters that range from sexuality to liberalism, from psychiatry to prisons, we can discern a persistent attention to the issues of power and knowledge as they relate to the formation and constitution of modern subjectivities. Foucault has been very influential in shaping our understanding of power, moving away from a mode of conceiving power as possessed by, or concentrated in the hands of, individual social actors, toward a mode of conceiving power as diffused, embodied, and enacted through various practices of discourse, knowledge, and what he himself called "regimes of truth" (Foucault 1980, 132). Foucault's view of power thus travels beyond the action-centric views of power (where power is seen as dispositions possessed and actions exercised by individuals) and even further than the institutional view of power (where individuals are differentially empowered and disempowered by discreet institutions), towards a systemic and constitutive view of power as pervasive, dispersed, and productive.

Our understanding of power, according to Foucault, has long been dominated by what he called "sovereign power", that is, repressive power which can become concentrated in the hands of a few and which can be exercised against others' will. What could account for the predominance of the sovereign view of power, Foucault suggests, is the fact that the modern discourses of power—that is, discourses which constituted the sovereign view of power—originated in an intimate proximity to the institution of the monarchy as it developed in the European Middle Ages, in the context of then commonly widespread struggles and conflicts among feudal barons and lords:

The monarchy presented itself as a referee, a power capable of putting an end to war, violence and pillage and saying no to these struggles and private feuds. It made itself acceptable by allocating itself a juridical and negative function, albeit one whose limits it naturally began at once to overstep. Sovereign, law and prohibition formed a system of representation of power which was extended during the subsequent era by the theories of right: political theory has never ceased to be obsessed with the person of the sovereign. (Foucault 1980, 121)

This sovereign view of power, described by Foucault also as "juridical and negative rather than technical and positive" (1980, 121) is not dissimilar to the episodic view of power, which can be found in the works of Weber, Dahl, and Lukes among others and which has held significant sway within Anglophone social and political theory and philosophy. As Clegg notes, "From its genesis in Hobbes to its maturation in Lukes, the concept of power is primarily of something which denies, forestalls, represses, prevents" (Clegg 1989, 156). We may indeed recall that some of the crucial forms of episodic power are coercion, manipulation and force as discussed in Chapter 2. Importantly, the sovereign view of power is characterised by the intermittent and episodic exercise of power. As Zygmunt Bauman (1982, 10) wrote, one of the main functions of sovereign power in the Middle Ages was the extraction of surplus product from peasant and villein producers. Since such extraction took place episodically, in accordance with the seasonal cycle of the agricultural production, sovereign power, too, was exercised episodically, in accordance with the seasonal extraction of the surplus product; and once the producers were coerced or forced to give away a (usually significant) fraction of their surplus product, they then could be left to their own devices (1982, 40). How the peasants and the serfs went about their daily lives were largely irrelevant to the sovereign monarch; what mattered was the task of ensuring the extraction of the surplus product through what Ernest Gellner called the "the dentistry state", a state engaged in the "business of extracting wealth by torture" (1979, 466). The sovereign monarch had thus no real interest in disciplining the body of the average peasant, with sovereign power having been exercised episodically.

The legitimacy of sovereign power was based on the claim that all power comes from God: in submitting to the rule of the monarch, the subjects were in effect submitting to the will of God (Ullmann 1965). This way of justifying sovereign power came to be known as the doctrine of the Divine Right of Kings, according to which subjects must obey and submit to the rule of their monarch, whose sovereign power is ordained by God (Figgis 1914, 5–6). It is in the context of this ecclesiastical justification of sovereign power that Hobbes, one of the originators of the action-centric view of power, proposed his distinctively secular and modern justification of sovereign power. As Hobbes (1839) imagined, in the state of nature, people have roughly equal powers. In pursuing their lives, individuals compete for scarce resources, leading to chronic competition. Their survival instinct drives them to preempt the attacks of others by attacking them first. The sense of uncertainty haunts even the strongest of individuals, because even they are vulnerable to attack, for example, when they sleep. The final outcome is a state of "war of every man against every man" (1839, 115), and life is "solitary, poor, nasty, brutish and short" (1839, 113). The state of nature is thus one of perpetual war. The solution, as proposed by Hobbes, is a social contract, whereby people confer all their powers to one man, thereby establishing sovereign power, whose purpose is to direct individual actions towards the common benefit, if necessary, through coercion or force. This Hobbesian mode of representing power appears to have largely survived in the current political theory (e.g., Kerny 2006, 288), which is still often preoccupied with the question of justification of state authority, for as Foucault argued, "political theory has never ceased to be obsessed with the person of the sovereign", while the political theories of right "still continue today to busy themselves with the problem of sovereignty" (Foucault 1980, 121).

Unsatisfied with this preoccupation with the sovereign view of power, Foucault put forward a rather radical and metaphorically gruesome proposition: "We need to cut off the King's head" (1980, 121). Remarkably, by cutting off the king's head, Foucault in fact kills two birds with one stone, so to speak, for his critique against the sovereign view of power appears to have had (at least) two noteworthy interconnected implications. Firstly, his argument emphasised that there is more to power than the power of the state. To be sure, Foucault does not mean to deny the importance of state power. Rather, his argument is that "relations of power, and hence the analysis that must be made of them, necessarily extend beyond the limits of the state" (1980, 122), not only because the state, with all its mighty administrative and military structures and apparatuses cannot occupy the entire domain of actual power relations, but also because the state, for Foucault, can only emerge on the foundation of other networks of power relations that exist in the broader human society. In his own words, the state "is superstructural in relation to a whole series of power networks that invest the body, sexuality, the family, kinship, knowledge, technology and so forth" (Foucault 1980, 122).

Secondly, Foucault's critique of the sovereign view of power questioned the claims that the concepts of action-centric views of power (i.e., episodic and dispositional concepts such as 'power-over' and 'power-to') are somehow analytically the most primitive or fundamental, since the individual who performs the actions in question is herself constituted or produced by power. An individual, for Foucault, should not be considered as "a sort of elementary nucleus, a primitive atom, a multiple and inert material on which power comes to fasten or against which it happens to strike, and in so doing subdues or crushes individuals", insofar as "it is already one of the prime effects of power that certain bodies, certain gestures, certain discourses, certain desires, come to be identified and constituted as individuals" (1980, 98). This echoes the point made by Spinoza about power: "[E]verything whereby a man is determined to act should be referred to the power of Nature insofar as this power is expressed through the nature of this or that man. For . . . he does nothing that is not in accordance with the laws and rules of Nature" (Spinoza 2002, 684). Thus, the individual, for Foucault, is "an effect of power, and at the same time, or precisely to the extent to which it is that effect, it is the element of its articulation. The individual which power has constituted is at the same time its vehicle" (Foucault 1980, 98). Hence, in the words of Clegg, "the episodic and agency view of power cannot be taken as any kind of analytic fundamental or primitive conception" (1989).

5.3. DISCIPLINARY POWER AND TECHNOLOGY

Foucault wrote that from the seventeenth century onwards, sovereign power had been overtaken by novel forms and regimes of social power, namely, a "capillary form" of power that "reaches into the very grain of individuals, touches their bodies and inserts itself into their actions and attitudes, their discourses, learning processes and everyday lives" and "a synaptic regime of power, a regime of its exercise within the social body, rather than from above it" (1980, 39). In *Discipline and Punish* (1977), Foucault viewed power as a technique or method that attains its strategic objective through its disciplinary nature, hence the notion of 'disciplinary power'. According to Foucault, the techniques, methods and practices of surveillance of individuals, which initially emerged as effective tools for disciplining individual subjects within administrative institutions such as the prison, became eventually widespread through other state institutions such as the school, the asylum, the army, and the factory. The disciplinary conception of power, expounded in *Discipline and Punish*, therefore applies not only to individual subjects such as criminals but also to a wide array of subjects who may show signs of deviance, delinquency, or abnormality of behaviour, subjects such as pupils, students, soldiers, and factory workers.

This disciplinary conception of power views power as constitutive and productive rather than possessed and exercised, by placing in the foreground the techniques, methods, measures, regimens, and regulations that come to serve in distinguishing the normal, the average, the sane, and the obedient from the deviant, the pathological, and the criminal. The techniques and methods implicated in disciplinary power aim to constitute and produce obedient social subjects capable of moral self-discipline. On this understanding, disciplinary power is not at the service of a sovereign individual or some dominant class, interest or ideology but is dispersed or diffused throughout various social, clinical, and carceral institutions, as well as systems and networks of social relations. By shifting the focus away from the sovereign conception of power, Foucault opens up new ways of seeing power which is at work within various social institutions and practices, while drawing our attention to the dialectic of power and resistance.

To explain the disciplinary character of modern power, Foucault employs the idea of the Panopticon, initially laid down by Jeremy Bentham, which is an architectural structure for a model prison that can "induce in the inmate a state of conscious and permanent visibility that assures the automatic functioning of power" (Foucault 1977, 201). The architectural structure of the Panopticon prison is described by Foucault as consisting of

> [a] perimeter building in the form of a ring. At the centre of this, a tower, pierced by large windows opening on to the inner face of the ring. The outer building is divided into cells each of which traverses the whole thickness of the building. These cells have two windows, one opening on to the inside, facing the windows of the central tower, the other, outer one allowing daylight to pass through the whole cell. All that is then needed is to put an overseer in the tower and place in each of the cells a lunatic, a patient, a convict, a worker or a schoolboy. The back lighting enables one to pick out from the central tower the little captive silhouettes in the ring of cells. In short, the principle of the dungeon is reversed; daylight and the overseer's gaze capture the inmate more effectively than darkness, which afforded after all a sort of protection. (Foucault 1980, 147)

The architectural design of the Panopticon entails that there is a one-directional monitoring or surveillance of the behaviour of the inmates; that is, while the guard can observe the behaviour of the prisoners in their cells, the prisoners themselves cannot see the guard in the tower. This means that for the Panopticon prison to have its powerful effect, the guard need not be in the tower all the time or ever at all; it is sufficient to make the prisoners believe that they are under constant surveillance in order to get them to modify and adjust their behaviour accordingly. The one-directional character of surveillance built into the structure of the Panopticon prison therefore has two important implications for the exercise of power in modern society. On the one hand, the exercise of power is disconnected from the figure or the person of the sovereign; power becomes impersonal. That is, by ensuring that "the surveillance is permanent in its effects, even if it is discontinuous in its action" (Foucault 1977, 201), the Panopticon prison transforms into "a machine for creating and sustaining a power relation independent of the person who exercises it" (1977, 201). On the other hand, the asymmetrical character of surveillance suggests that both the means and the object of power converge, to some extent, in the body of the individual subject of power; as the inmate, according to Foucault, becomes "caught up in a power situation of which they are themselves the bearers" (1977, 201).

The principle of the Panopticon suggests that in producing obedient and docile individual subjects, it is not always necessary to subject them to physical force or constraints, given that the strategic objective can be achieved simply by placing these individuals under surveillance. Compare, for example, the uses of the speed bump and the speed camera for slowing down speeding vehicles on roads (see also Subsection 2.3.2). The speed bump slows down speeding drivers by sheer physical force: driving over the speed bump at an accelerated speed would normally cause a strong physical impact that can harm the physical structure of the speeding vehicle, and drivers normally prefer to avoid causing such harm to their vehicles, especially when the latter is privately owned. The speed camera, on the other hand, can achieve the same effect, not by physical force but by assuring the potentially deviant drivers that if they violate the speeding restrictions, their number plates can be photographically registered and transmitted to the law enforcement agency, who can later use this information to identify and penalise the violator. In this example, the speed camera is largely based on the same principle as the Panopticon prison: it induces in potentially deviant drivers a specific state of consciousness which assures the automatic functioning of power.

The idea of 'panopticon effect' can also shed light on how disciplinary power relations can extend to individual and social behaviour online. Since disciplinary power comprises techniques which aim to produce morally obedient individuals capable of self-discipline, it is also possible to replicate such effects with regard to the uses of information and communication networks. The extension of disciplinary power into information and communication networks, such as the internet, can both curtail and direct specific uses of the internet by users, citizens, and other entities. To understand this, let us consider some of the practices of surveillance and censorship on the internet.

While censorship and surveillance are often treated as distinct and separate phenomena, there is an increasing convergence between the two practices in the current digital age. As Cory Doctorow (2012), journalist and the author of the novel *Little Brother* (2008), writes, in the pre-digital age, censorship of publications, such as *Ulysses* by James Joyce in Britain in the 1920s and 1930s, would mainly take the form of imposing a ban on the sales of copies of the publication to be censored. In the age of the internet, however, preventing people from accessing censored information on the World Wide Web, "the national censor wall must intercept all your outgoing internet requests and examine them to determine whether they are for the banned website" (Doctorow 2012). This, according to Doctorow, is the main difference between old and new form of censorship: "Today, censorship is inseparable from surveillance".

Importantly, the convergence between the practices of censorship and surveillance, especially in particular contexts where state and government agencies threaten internet users with penalties for accessing 'wrong' parts of the internet, can lead to 'self-censorship', a term used to refer to "the controlling of what one says or does in order to avoid annoying or offending others but without being told officially that such control is necessary" (Clark & Grech 2017). The combination of the visibility of individual behaviour online and the threat of punishment for accessing prohibited information online has the disciplinary effect of producing self-censoring internet users. Depending on the particular jurisdiction, the threat of punishment can be either explicit or implicit, with the form of a punishment itself ranging from mild (e.g., paying a fine) to severe (e.g., imprisonment, torture, or death).

The extension of the principle of panopticism to information and communication networks can result in self-censorship, as people living in authoritarian states characterised by their active surveillance of online behaviour learn to behave in ways that do not arouse the interest or the suspicion of surveillance agencies. The so-called 'chilling effects' on individual behaviour online often result from one's fear of potential reprisals and penalties and other factors that constrain free expression. George Orwell once noted that "the sinister fact about literary censorship in England is that it is largely voluntary" (quoted in Sandberg 2012). Thus, self-censorship largely results from the internalisation of fears and assumptions about potential negative implications and punitive consequences of particular online behaviour in contexts characterised by the presence of surveillance systems. Self-censorship, as Sandberg (2012) writes, is "often based on little more than assumptions".

We may, in fact, need a new concept to refer to the increasing convergence of censorship and surveillance, perhaps a conjoint term, such as 'censorship/surveillance' (in reference to one of Foucault's own conjoint terms, 'power/knowledge'). The strategic convergence of censorship and surveillance furthermore suggests the convergence of the means and ends of power in the body of the individual subject of power. Due to the panopticism of modern power, as Foucault writes, the modern subjects of power become "caught up in a power situation of which they are themselves the bearers" (1977, 201). The panopticon effect online entails that the person, or persons, monitored become both the means and the object of power, perhaps with as little intervention as possible, so long as the persons monitored are made aware that they are under the watchful gaze of state or government surveillance agencies.

The disciplinary character of such modern forms of surveillance and censorship can be overt, as can be seen, for example, in the case of Saudi Arabia. A noteworthy aspect of the Saudi censorship scheme, according to Morozov (2011, 104), is that users attempting to access prohibited websites and content are often served with an explanation of the 'reason' why the content is being blocked. Specifically, in the case of banned pornographic sites, users are presented with a notice explaining the reason for the ban, which, remarkably, cites an academic article on pornography authored by Cass Sunstein (1986), an eminent legal scholar (Morozov 2011). The practice of offering reasons for the prohibition of certain content online suggests that users become aware of the fact that their activities are being actively monitored, which in turn can lead to self-regulation of online behaviour in some individuals. Another self-disciplinary feature of the Saudi censorship scheme is that it encourages the citizens to report any links to online content they may find offensive, with more than a thousand of such reports being submitted to the Communications and Information Technology Commission every day (Morozov 2011). The practice of reporting 'offensive' content can be regarded as the outsourcing of part of the state surveillance activities to ordinary citizens, which is another case of the convergence of means and objects of power, this time in the population as whole.

5.4. BIOPOWER AND TECHNOLOGY

Disciplinary power, discussed in the previous section, is one of two distinct forms of power which, for Foucault, began to exemplify the modern period. The other form of power Foucault calls "bio-power" (1978, 140–141). While the strategic effects of disciplinary power have individuals as their object, biopower is directed toward the control and regulation of human population as a whole. In other words, disciplinary power is "micropolitical", whereas biopower is "macropolitical" (Kelly 2009, 43). While both disciplinary power and biopower, for Foucault, stand in contrast to the sovereign conception of power, disciplinary power and biopower "complement one another without conflict", since the two modern technologies of power operate at different, micro and macro, levels of politics and society (Kelly 2009). According to Kelly (2009), disciplinary power and biopower have not so much replaced sovereign power, but have supplemented it. Biopower, for example, can coexist together with sovereign power. What separates "those who are subject to the lethal technology of sovereignty, namely criminals, proscribed ethnic groups, . . . foreigners, and those who must be 'made to live' by biopower" is, for Foucault, racism (Foucault 2003, 254–256; see also Kelly 2004).

The emergence of biopower, according to Foucault, can be observed in the shift in the conduct of warfare: "Wars are no longer waged in the name of a sovereign who must be defended; they are waged on behalf of the existence of everyone; entire populations are mobilized for the purpose of wholesale slaughter in the name of life necessity" (1978, 137). The chief concern of biopower therefore is not so much the individual but the human population as a whole: "It is as managers of life and survival, of bodies and the race, that so many regimes have been able to wage so many wars, causing so many men to be killed" (1978, 137). Nonetheless, the concept of biopower as discussed by Foucault is regarded as both rather broad and vague (e.g., Rose 2007, 54), an issue apparently familiar to many students of Foucault (e.g., Harvey 2008). Foucault himself discussed the concept in his lecture delivered at the Collège de France (Foucault 2003, 239–264) and in a chapter titled 'Right of Death and Power over Life' in his book *La Volonté de Savoir* (1976; English translation Foucault 1978). Foucault appears to have also intended to elaborate on the concept in one of the six proposed volumes of the history of sexuality, the titles of which appeared on the back cover of the book. However, the task of further elaborating the concept remained unfulfilled. While the concept of biopower was subsequently taken up by a number of authors (e.g., Deleuze 1988; Brenner 2000; Hardt & Negri 2000), Rabinow and Rose note that at the dawn of the current "biological century", the term 'biopower' was more likely to be regarded as a term denoting "the generation of energy from renewable biological material" (2006, 197), instead of a contested field of issues and strategies crucial for understanding of the political aspects of the biology of human life.

In this context, *The Politics of Life Itself* by Nikolas Rose (2007) has been a very welcome intellectual contribution that added much needed substance to Foucault's original formulation of biopower. Rose offers an accessible interpretation of the concept, especially of what biopower and biopolitics has come to mean in the current century marked by fundamental scientific and technological advances in the life sciences, which well complements the original writings of Foucault. Rose begins the discussion by pointing out that since the initial articulation of the concept by Foucault, there has been a major transformation of the technological capacity of biopower. Therefore, Rose argues, biopower in its original formulation by Foucault, while retaining considerable analytical utility, is insufficient for explaining the moral, social, and political consequences of new biomedical technologies that have been taking shape since the last quarter of the last century (Rose 2007, 9–10). The main argument advanced by Rose is that in the current century, biopower and biopolitics have been undergoing transformation along five "pathways" of (1) *molecularization*, (2) *optimization*, (3) *subjectification*, (4) *somatic expertise*, and (5) *economies of vitality*. These transformations, for Rose, have resulted in "an emergent form of life", where "biopolitics has become inextricably intertwined with bioeconomics" (2007, 7). Let us consider each of these five pathways of the transformation of biopower and biopolitics in the twenty-first century and the role technology has played within these transformations.

The 'molecularization' of life, according to Rose, marks the shift from visualising life at the 'molar' level, which is the tangible and visible level of limbs, organs, and tissues, to visualising life at the molecular level, which is the level of molecules, stem cells, nucleotides, spermatozoa, chromosomes, and so forth. The molecularization of life, for Rose, entails a new "style of thought" (Fleck 1979; cited in Rose 2007, 12), which is a particular way of thinking, seeing, and practicing that involves "formulating statements that are only possible and intelligible within that way of thinking". Such a new style of thought, Rose argues, has emerged in the biological and medical sciences, modifying their objects of inquiry such that their appearances, properties, and relations can be formulated at the molecular level. The emergence of seeing and acting upon life at the molecular level, according to Rose, has been made possible through the invention and commodification of "new technologies of visualization", screening devices which allow users to peer into the interior of the human body: mammograms, ultrasound, foetal images, EEG traces, PET, SPECT, fMRI scans, and so on (Rose 2007, 14). In this technological context, the laboratory has been transformed into "a kind of factory for the creation of new forms of molecular life", giving rise to forms of "molecular biopolitics", where miniscule elements of life can be mobilised, recombined, and controlled, thereby creating new ways of manipulating and politicising the very building blocks of life (Rose 2007, 13, 15). Politics, in other words, reaches the microscopic depths hitherto unseen.

Biopolitics has also been transformed with the emergence of what Rose calls "technologies of optimization" (2007, 15–16). These technologies, for Rose, do not simply aim to cure disease but seek to change the very nature of biological organisms, through configuration of the vital biological processes with the aim of maximising their functioning. The main characteristic of the technologies of optimisation, according to Rose, is their forward-looking vision, insofar as they "seek to reshape the vital future by action in the vital present" (2007, 18). On the one hand, there are efforts to assess, identify and treat certain genomic traits in persons in the present with regard to potential health conditions they may endure in the future. Such efforts are directed at the individual susceptibility to those disorders that may have an underlying genetic cause; hence, there is an ongoing search for genomic variations that can increase susceptibility to certain disorders and diseases, accompanied by the development of genetic tests for embryos, foetuses, children and adults. On the other hand, there have emerged new avenues for human enhancement. To be sure, the practice of human enhancement is not at all new: humans have invariably aimed to enhance our bodily and intellectual capabilities through "prayer, meditation, diet, spells, physical and spiritual exercises" (Rose 2007, 20). What is new, according to Rose, is the sense of novelty and disquiet arising from beliefs that new practices of enhancement have become more "powerful, precise, targeted, and successful" (2007, 20). In other words, there have been a shift "from normalization to customization" (Clarke et al. 2003, 181; cited in Rose 2007, 20).

Another facet of biopolitics in the twenty-first century is the role biology and medicine play in the shaping of individual subjectivities. Biology, for example, has long had a role in the formation of individual and group subjectivities, where racial, national, and ethnic differences have, at least in part, been represented on the basis of their biological characteristics. Today, according to Rose, we are witnessing the emergence of novel forms of biomedical subjectification. For example, Rose cites the case study conducted by Adriana Petryna (2002), who has explored how Ukrainian citizens, following the Chernobyl nuclear catastrophe, made claims and demands for compensation and the redistribution of resources on the basis of their damaged biology. In a similar vein, Paul Rabinow (1996) introduced the notion of "biosociality" to characterise novel forms of group identification in terms of genomics. Similar forms of biomedical subjectification have been identified by Rose, in collaboration with Carlos Novas (Rose and Novas 2004). The latter two authors have introduced the concept of "biological citizenship" to specify the ways in which ideas and practices of "citizenship has been shaped by conceptions of the specific vital characteristics of human beings and has been the target of medical practices since at least the eighteenth century in the West" (Rose 2007).

Developments concerning what Rose has called "somatic expertise" are giving rise to novel forms of expert power that have come to govern specific aspects of the bodily existence of individuals and groups (2007, 6, 9, 27–31). These developments involve the emergence of various scientists, technologists, ethics experts, and other professionals who claim expertise in managing particular aspects of bodily human existence. The emergence of these new forms of expertise and biopower are inextricably connected to the complexity of modern technologies of life. The more technologies can uncover potential threats to life and health, the more there is need for what Rose, following Foucault (1982, 782), describes as "pastoral power" of the experts, whose function is to guide, care, support, and advise individuals and their families as they face personal, medical, or ethical issues (2007, 73–76). According to Rose, some of the most remarkable among such professionals and experts are geneticists, who can help individuals manage and navigate the ever shifting and complex landscape of new genetic medicine, treatments, and technologies, as well as different kinds of ethics experts, such as *bio*-ethicists and *gen*-ethicists, who claim capacity and expertise in evaluating and adjudicating right and wrong, just and unjust, forms of somatic activity, expertise, and power. These experts, as Rose notes, have even been "enrolled in the government and legitimation of biomedical practices from bench to clinic and marketplace" (Rose 2007, 6).

Finally, a fifth kind of transformation that biopower and biopolitics have been undergoing in the past decades concerns the ever-growing connection between biology and economy. According to Rose, new connections have sprung up between truth and value, vitality and capital (2007, 6–7). Prompted by the search for capital and value, there has been an increasing convergence between, on the one hand, the demands for corporate shareholder value and, on the other, the human and social value, the latter being invested in the hope for cure and prevention. These developments have given rise to a new economic space—"bioeconomy"; and to a new form of capital—"biocapital" (2007, 252–258). In this new political and economic environment, there has been a transformation of the networks and relationships between pharmaceutical companies, science and research institutions, and stock markets. "Life itself", as Rose argues, "has been made amenable to these new economic relations, as vitality is decomposed into a series of distinct and discrete objects—that can be isolated, delimited, stored, accumulated, mobilized, and exchanged, accorded a discrete value, traded across time, space, species, contexts, enterprises—in the service of many distinct objectives" (2007, 7).

5.5. CONCLUSION

What are some of the implications of the elimination of the sovereign power as advocated by Foucault? First, the views of power that take intentions and intentionality as central to defining power (e.g., Dahl & Lindblom 1953; Morriss 2002) would be reconsidered. As we have seen in the example of technology design, in considering the incorporation of values and needs of particular users (especially those traditionally undervalued or underrepresented) in the design of the technology, the issues of power structures and institutions should be accorded a more prominent place (in addition to the intentions and interests of technology designers). Second, the asymmetrical and impositional interpretations of power, whereby some individuals and groups can impose their will on others in asymmetrical social relations, will no longer dominate our understanding of power, instead leading us to consider the self-imposed character of power and power relations. This would also entail the recognition that there is more to power than its sovereign/episodic conception. As our discussion of disciplinary power and biopower suggests, technology can become implicated in power relations which are not subject to one controlling entity (such as the state) but are impersonal, dispersed, and systemic. Third, the kind of singularism about the concept of power which tended to dominate the main strands of Anglo-American scholarship on power is bound to be infused with what Clegg described as "a greater modesty in the ease with which a single concept of power could be extemporized for all purposes" (1989, 186). In this particular respect, the poststructuralist approach to power lends further support to the pluralist approach adopted in this book. Somewhat dogmatic attitudes frequently underlying the attempts to find the single best definition of power is very likely to blind us to the various ways in which technology is implicated in the working of power in our contemporary society. Such attempts can result in making technological power more insidious and invisible, thereby making it harder to counter and resist. Eyes trained to see power in action-centric, or even in terms of discrete institutions, might not be able to see other forms of power at work, forms of power that define and constitute our very selves. And, on this note, we can end the first part of the book.

Part II

Power and Ethics of Technology

The second part of the book concerns itself with the role the concept of power plays in the ethics of technology. The following chapters (chapters 6, 7, 8, and 9) offer a provisional groundwork for elucidating the role of the concepts of power and technological power in ethical analyses of technology. Chapter 6 applies the theoretical framework developed in the first part of the book to the emerging field of the ethics of algorithms. Chapter 7 explores the conceptual connection of the concept of power to the ethical concepts of responsibility, vulnerability, authenticity, and trust. Chapter 8 considers the practical implications of power for the implementation of responsible research and innovation in Europe. Finally, Chapter 9 considers implications of power to technological innovation in the absence of ethical and democratic institutions and mechanisms for guiding technological innovation processes in society. To do proper justice to the question of power in the ethics of technology is undoubtedly beyond the scope of the proposed book. However, it is hoped that the preliminary inquiry presented in this part of the book can be used as a stepping-stone for future research in this area.

Chapter Six

Power in the Ethics of Algorithms

This chapter begins the second part of the book that concerns the role of the concept of power in the ethics of technology. The chapter commences this philosophical exploration by considering the role of the concept of power in the nascent field of the ethics of algorithms. The particular focus on the ethics of algorithms is warranted for two main reasons. First, the ethics of algorithms is an emerging area of ethical inquiry, within the broader field of the ethics of technology, to which we can apply, as a test case, the philosophical framework of power developed in the first part of the book. Second, the ethics of algorithms is that area of ethical inquiry which appears to have been responsible for reviving interest in the role of the concept of power in the ethical and critical discourses on technology. By applying the framework of power, developed in chapters 1–5, to the domain of computational and machine-learning algorithms, the present chapter aims to give an account of issues of power and prejudice resulting from various uses of algorithms in web-based applications, intelligent machines, and data analytics systems. Overall, the chapter aims to show that a majority of emergent ethical issues with algorithms can be identified by considering different effects of algorithms on the construction, distribution, and exercise of power in society.

6.1. ALGORITHMS, POWER, AND BIAS

Whether embedded in personal computers, intelligent machines, or data analytics systems, algorithms have become akin to a digital Swiss army knife, with an almost endless number of possibilities and uses to which they can be applied. They are claimed to be able to produce expensive works of art (Holmes 2018), rival doctors in diagnosing lung diseases (Lay 2018), help social services identify children at risk (Hurst 2018), provide "an Eton education for all" schoolchildren (Burgess 2018), assist spies and security services in counterintelligence activities (Baker 2018), and even take charge of cannabis farms, allowing their owners to 'chill out' (Bridge 2018). Most remarkably, however, just like people, algorithms can become powerful and prejudiced. While we cannot, for obvious reasons of space, explore each and every characteristic of algorithms mentioned above, our attention will be directed at the notions of power and prejudice frequently attributed to algorithms.

That algorithms can be powerful has been the focus of a growing number of recent philosophical and critical studies, especially within the media and communication studies (e.g., Lash 2007; Beer 2009; Golumbia 2009; Cheney-Lippold 2011; Mager 2012; Snake-Beings 2013; Pasquale 2015; Bucher 2012, 2016; Danaher 2016; Sattarov 2016; Yeung 2017). Within these works, algorithms are seen as possessing power and as playing a role in the exercise and distribution of power in society. For example, it has been argued that "power is increasingly in the algorithm" (Lash 2007, 71) and that "algorithms have the capacity to shape social and cultural formations and impact directly on individual lives" (Beer 2009, 994). Such concerns about the power of algorithms has led some authors to variously emphasise the importance of adopting critical approaches in governing and regulating algorithms (e.g., Diakopoulos 2014; Neyland & Möllers 2016; Kitchin 2016). However, despite the growing interest in the notion of algorithmic power, little effort has been made to distinguish between different senses of algorithmic power. As the chapters in the first part of the book make it clear, there is more than one sense of power, which range from action-centric to structure-centric views (see also, e.g., Saar 2010). Despite the ongoing debates, power is a fairly well-researched concept, with different conceptualisations of power having been developed within social and political theory. Hence, the first of the two main goals of this chapter is to elucidate how algorithms relate to these different senses of power extant in the social and political theory. By doing so, the discussion aims to bring some analytical rigour and a level of interdisciplinarity to facilitate cross-disciplinary communication between critical studies of algorithms, social and political science, and philosophy.

In addition to being powerful, algorithms are also frequently noted for their potential to be prejudiced or biased against, or in favour of, certain individuals, groups, processes, or outcomes. The biased and prejudicial character of algorithms has been at the centre of attention of a number of scholarly inquires, especially within the ethics of information technologies (e.g., Friedman & Nissenbaum 1996; Brey 1998; Kraemer et al. 2011; Ananny 2016; Hajian et al. 2016; Mittelstadt & Floridi 2016; Mittelstadt et al. 2016; O'Neil 2016; Mann & O'Neil 2016; Sattarov & Coeckelbergh 2017). However, what appears to be missing from these studies of prejudiced algorithms is an account of the relationship between biased algorithms and algorithmic power. At a first glance, there seems to be some kind of relationship between bias and power in general. For example, within some social system, one group can exercise power over another group by excluding them from the distribution of economic resources, benefits, or other relevant social goods through what Eric Schattsneider has called "mobilization of bias" (1960). While acknowledging this connection between bias and power in general, it still remains to be explicated how this relationship between bias and power plays out in the case of algorithms. This constitutes the second of the two main goals of our discussion in this chapter.

6.2. ALGORITHMIC POWER

Before we can sensibly speak about algorithmic power, we should first of all explicate (1) what we mean by the terms 'algorithm' and 'power'. In current literature, many critical studies of algorithms often fail on both counts: they do not proffer a formal or technical definition of algorithms (Hill 2015), and they do not explicate their conception of power (Sattarov & Coeckelbergh 2017). These shortcomings are also applicable to popular discourses of algorithms (Hill 2015). To avoid such shortcomings, let us first define the meanings of these terms as they are employed in this chapter.

Etymologically, the term 'algorithm' traces its root to the name of Al-Khwārizmī (circa 780–850), an influential mathematician, whose Arabic name means "the native of Khwārizm", a region in modern-day Uzbekistan. The term 'algorithm' derived from the Latinisation of Al-Khwārizmī's name, in the course of translating his treatise on Hindu numerals into Latin as *Algoritmi de Numero Indorum* (e.g., Corona 2006). In modern English language, the term has acquired the meaning of "a set of rules or procedures that must be followed in solving a particular problem" (*Oxford English Dictionary* 2010). In current philosophical and sociological literature, it is possible to distinguish between the *narrower* (or technical) and *broader* (or popular) senses of the term 'algorithm'. In its narrower sense, an algorithm can be defined as "a finite, abstract, effective, compound control structure, imperatively given, accomplishing a given purpose under given provisions" (Hill 2015, 47). In its broader sense, algorithms can refer "not just to a set of instructions sufficiently precise that they can be encoded in a computer program but also to that program running on a physical computer system and having effects on other systems" (MacKenzie 2018, 1637).

The discussion here will mainly follow the broader sense, although on occasions it will also rely on the narrower sense (where it will be specified). The adoption of the broader sense is partly due to the fact that this sense is often implicit in many critical studies and popular discourses of algorithms (e.g., Burrell 2016; Kitchin 2016; MacKenzie 2018). The adoption of the broader sense is also required given that algorithms, as sets of precise instructions, acquire the power they have when embedded and operationalised within certain software and hardware systems (although it is possible, under certain provisos, as shown below (Subsection 6.2.2), to attribute powerful qualities to algorithms understood in the narrower sense as a set of precise instructions).

As for the definition of power, the first part of the book has identified four main senses of power, such as episodic, dispositional, systemic, and constitutive. To reiterate briefly, on the episodic view, power is conceptualised as a relationship in which one actor exercises power over the other, for example, by means of seduction, coercion, manipulation, and the like. On the dispositional view, power is regarded as a capacity, ability, or potential of a person or entity to bring about certain relevant outcomes. On the systemic view, power is understood to be a property of various social, economic, and political institutions that structurally create possibilities for individual action. Finally, on the constitutive view, power is seen as systemically constituting, or producing, social actors themselves. The following subsections aim to interpret algorithmic power based on these four perspectives. They thus aim to sort out different senses in which algorithms can be described as powerful.

6.2.1. The Episodic View of Algorithmic Power

When viewed from the episodic perspective, algorithmic power can refer to the ways in which algorithms can affect the behaviour of users through various kinds of incentives and motivations, whether overtly or covertly. For instance, consider the following two studies of Facebook algorithms, which suggest that algorithms can affect the moods of users and direct them toward certain kinds of actions. A recent study published in the *Proceedings of the National Academy of the Sciences* (PNAS) presented "experimental evidence for massive-scale contagion via social networks" by tweaking the number of positive or negative posts shown to Facebook users by experimentally manipulating users' algorithmically managed 'News Feed' (Kramer et al. 2014). This algorithmically run feature of Facebook makes 'decisions' regarding which of the other users' "status updates" or news items a Facebook user sees on his or her page. The study purported to show that such news and updates can influence the mood and feelings of Facebook users as well as the character of their subsequent posts.

Another recent study of the social-psychological effects of algorithms conducted by Bucher (2016; see also Bucher 2012) suggests how Facebook users' particular "ways of thinking about what algorithms are" can be "productive of different moods and sensations" in those same users (2016, 32, 41). Bucher further argues that studying how algorithms can make people feel is crucial for understanding their social power (by which she evidently means the episodic power as discussed in Chapter 2). What these studies share in common is that they show how web algorithms can affect the moods and feelings of their users and can direct them toward certain kinds of actions. The crucial notion here is the production of specific guiding and motivating perceptions, moods, and ideas in the minds of those over whom power is to be exercised. This idea has been central in the social-psychological studies of forms of power conducted by French and Raven (1959).

Algorithms also play a role in persuasive power relations. Consider, algorithms underpinning so-called persuasive technologies—technologies defined as interactive computing systems designed to change the attitudes and behaviours of their users (Fogg 2003). Indeed, there has been a proliferation of algorithmic software systems designed to alter the behaviour of people through persuasion. Websites such as Amazon.com and eBay do not simply process purchases and transactions but also aim to persuade users to buy more products. For example, so-called algorithmic recommendation systems (e.g., Beer 2013, 63) make suggestions to users concerning what products or services to purchase based on the preferences exhibited, and other information provided, about these users. Such systems can also use the information and preferences exhibited by users on their previous visits to the webpage and the feedback generated from other users' comments about the services and products offered. Moreover, these recommendations can go beyond purchasing behaviour by giving users directions about when to exercise, which route to take, or who to contact, and so on (de Vries 2010, 81). The toolkit of persuasive techniques can benefit from greater powers of prediction produced through techniques of machine learning algorithms (e.g., Mackenzie 2015). Beyond the web marketing, persuasive algorithmic applications are spreading to areas such as education, environment, and healthcare.

6.2.2. The Dispositional View of Algorithmic Power

Dispositional power, as discussed in Chapter 3, conceives of power as ability, capacity, or potential. In Section 3.2, we further distinguished dispositional powers into human dispositional powers, such as abilities (e.g., one's ability to write and read), and artefactual dispositional powers, such as capacities (e.g., capacity of an engine to move a vehicle). The main difference between these two kinds of dispositional powers is that the former can be exercised at will. Indeed, this difference somewhat corresponds to that drawn by Aristotle between natural powers, such as the power of water to extinguish fire, and rational powers, such as one's ability to speak Latin. According to Aristotle, if the necessary and sufficient conditions for the actualisation of a natural power are satisfied, then the natural power is necessarily exercised: if you throw a burning cigarette into a cup of water, it will be extinguished. Yet, a rational power, for Aristotle, was a power that can be exercised at will: if a person is presented with the necessary conditions for exercising its rational power, he or she might choose not to exercise that power (Aristotle, *Theta* 1046a–1048a). The distinction drawn by Aristotle suggests that while algorithms cannot be ascribed 'rational powers', they can be attributed 'natural powers', such as the capacity of an engine to rotate.

What does this mean for algorithms? Consider, a piece of software containing a set of natural language processing algorithms for machine translation of Dutch speech into English. This set of algorithms can be said to possess the dispositional property, potential, or capacity to translate Dutch into English. Certainly, as a set of coded instructions, these algorithms cannot, by themselves, successfully translate Dutch into English, since a successful translation would require, among other things, an appropriate hardware, a competent operator, and so forth. (If, for example, they were spray-painted in the form of algorithmic pseudocode on the white cliffs of Dover, they would not exhibit much power perhaps apart from irritating a few locals.) However, even in the absence of such requirements, indeed in the absence of their actualisation, it still remains possible to say that these algorithms have a dispositional property, potential, or capacity to translate Dutch into English. The existence of this dispositional property becomes obvious, for example, when we distinguish this set of algorithms from some other set of algorithms designed for translating Sanskrit into Esperanto. Thus, it seems that, when Lash claims that "power is increasingly in the algorithm" (2007, 71), this claim should not be interpreted as suggesting that algorithms are becoming autonomous holders of rational powers which they can exercise whenever they please, but should be interpreted instead as meaning that in pursuit of power, algorithmic capacities are becoming a prominent means of acquiring power and exercising control.

6.2.3. The Systemic View of Algorithmic Power

While the episodic and dispositional views of power draw attention to algorithms themselves, the systemic perspective on algorithmic power places emphasis on the institutional and organisational structures in which algorithms are embedded. The consequences of algorithmic power would be seen as stemming from the wider socio-technical and institutional context, where algorithms are only an element among other elements bound together through social, political, and other institutions, processes, and practices. Put differently, the systemic view of power tends to shift attention from the technical sense of an algorithm as a precise set of mathematical instructions toward the broader sense of algorithms as a set of instructions embedded into software and hardware systems to perform a certain task or action, while these systems themselves form part of the broader social, economic, and political institutional reality.

In our current digital age, institutions are increasingly becoming permeated by algorithmic technologies and practices. Within institutions, actions and tasks previously performed by persons are increasingly delegated to algorithms that mediate social interactions, economic transactions, and political decisions (e.g., Floridi 2012; Portmess & Tower 2015). On this view algorithms are increasingly constituting social, economic, political, and other institutions. As argued in Chapter 4, there is a bidirectional relationship between institutions of power and technology. In one direction, institutions of power affect technology, for example, by setting the trajectory for technological innovation and development (Halperin et al. 2005), by shaping emerging technologies (Brey 2017), or altogether resisting and effectively blocking technological innovation and development (Acemoglu & Robinson 2000). In the opposite direction, technology can affect institution of power. For example, institutions consist of relatively stable relationships and networks of people and power. The relative endurance and stability of such networks require stable networks of communication which increasingly depend on information and communication technologies (Castells 2009).

This relationship of mutual influence is also applicable to algorithmic technologies, systems and practices. One the one hand, algorithms can affect institutions. For example, sorting and profiling algorithms can result in ethically unacceptable outcomes as observed, for example, in the case of discrimination against marginalised groups (Barocas & Selbst 2016), or in the case of online advertisements being targeted at users on the basis of their perceived ethnic origin (Sweeney 2013). These are cases of algorithms affecting institutions, whereby a specific algorithmic technology and practice contributes to the perpetuation of institutionalised forms of discrimination and racism. On the other hand, there are cases in which this relationship is reversed, where institutions impact or shape algorithmic technologies and practices. For instance, the General Data Protection Regulation of the European Union, which came into force in May 2018, is expected to have some implications for how algorithms are used in processes of automated decision-making, since the regulation (arguably) creates a legal right "whereby a user can ask for an explanation of an algorithmic decision that was made about them" (Goodman & Flaxman 2017, 50).[1] This is a case of institutions affecting algorithms, whereby political and legal institutions impact the specific applications of algorithms in social affairs.

6.2.4. The Constitutive View of Algorithmic Power

The *constitutive* view of power is very similar to the *institutional* view discussed above, insofar as the two conceptions regard power as inhering in the social, material, and institutional context or system that makes individual action possible (hence, they are together described as 'system-centric').

However, what makes the two views distinct is that the constitutive view goes a little further than the institutional view and makes the stronger claim that power not only makes individual action possible but also constitutes social actors themselves (see also Allen 2016). The constitutive view of power draws attention to the ways in which power shapes us as selves. It recognises power in its most pervasive aspect. From this perspective, technologies become implicated in various practices and processes that create identities and mould subjectivities.

To see how algorithms become implicated in such constitutive practices and processes, consider how different trading practices in stock and financial markets entailed different conceptions or ideals of being a good and efficient trader. In the broader context of financial market trading, it is possible to distinguish three main types of traders: (1) *floor traders* (also known as 'pit traders') who carry out trade transactions on the trading floor (or the pit) of a stock or commodities exchange, by using the method of 'open outcry' which involves shouting and hand gestures to communicate information about buy and sell orders (e.g., Zaloom 2006); (2) *screen traders* (also known as 'click traders') who execute their trade transactions via computer screens by clicking a mouse (e.g., Knorr Cetina & Bruegger 2002); and (3) *high-frequency traders* (HF traders) who develop algorithms used for high-frequency trading (as discussed in the previous subsection), and monitor the performance of these algorithms in the course of HF trading (e.g., MacKenzie 2018).

While these trading practices arose as a result of different institutional factors and technological advances, such as the computerisation and algorithmisation of trading practices (e.g., Currie & Lagoarde-Segot 2017), they also gave rise to different conceptions of what kind of skills, competences, or other human characteristics are valuable for different trading practices. For example, floor-trading practices required traders to have certain physical and vocal attributes that would make them hard to be ignored and more likely to be noticed on the trading floor (Borch & Lange 2016). With transition to screen trading, there emerged a need for traders with non-physical attributes, such as mathematics and computer skills (Zaloom 2006, 84). Finally, the emergence of high frequency trading further increased the demand for traders with computer skills, who are expected to be "very computer proficient in terms of using analysis language", "able to code", and "savant-like type" (Borch & Lange 2016, 10). Indeed, in their study of the effects of high-frequency trading practices on trader subjectivity, Borch and Lange (2016) observed that a third of the high-frequency traders they interviewed for their study had PhD degrees in areas such as mathematics, physics, and engineering (2016, 11). Consequently, highly educated high-frequency traders, according to Borch and Langue, bring a "research-based ethos" and "a highly disciplined work ethic" into their trading (2016, 11), which seems to be at odds with the lax attitude toward working hours characteristic of screen traders as observed by Zaloom (2006, 87–88).

Borch and Lange's study has also found that high-frequency traders regard emotions as something they should avoid as much as possible: "The calm, disciplined, almost machine-like approach of HF traders constitutes a bodily enactment of the ways in which computer algorithms are believed to be a rational response to . . . the kinds of errors that humans are purportedly prone to make" (2016, 13). Importantly, Borch and Lange document a number of self-disciplining techniques employed by high-frequency traders with the aim of curbing and suppressing their emotions which may interfere with the allegedly unerring performance of their algorithms. Such self-imposed regulatory techniques include creating complexity as a deliberate technique for keeping emotions at bay, maintaining a detached and impersonal attitude toward the algorithms they have created, replacing intuition with strict calculation, and adopting prudent, systematic, and research-based approach to taking risks (Borch & Lange 2016, 16–19). The crucial point here is that these techniques are what Foucault called 'techniques of the self': "techniques that permit individuals to effect, by their own means, a certain number of operations on their own bodies, their own souls, their own thoughts, their own conduct, and this in a manner so as to transform themselves, modify themselves, and to attain a certain state of perfection, happiness, purity, supernatural power" (Foucault 1997, 177). On this view, the formation and production of subjects of power occurs when individuals subject themselves to various techniques of the self. Importantly, the self-imposed nature of such disciplining practices is what makes power constitutive.

6.3. ALGORITHMIC BIAS

Having elaborated on the four main senses of algorithmic power, the present section takes the discussion further by considering cases of so-called 'biased' algorithms. An examination of the biased character of algorithms can shed further light on the systemic and institutional views of algorithmic power, insofar as biased algorithms can contribute to institutional forms of bias and prejudice, such as institutional racism and/or sexism. The discussion in this section is divided into three parts. First, we discuss the kinds of biased algorithms distinguished on the basis of their *object*, that is, who or what is being subjected to bias, prejudice, or discrimination (e.g., social bias, user bias, informational bias). Second, we discuss the kinds of biased algorithms distinguished on the basis of their *origin*, that is, how this or that bias in algorithms has come about (e.g., as a result of intentional design, or as an unintended consequence, etc.). Finally, we examine how *bias* in algorithms relates to *power* in algorithms, that is, in what way biased algorithms can be considered to be also powerful.

6.3.1. Objects of Algorithmic Bias

Not all instances of biased algorithms are the same. It is in fact possible to distinguish different kinds of biased algorithms on the basis of the objects or targets of these biased algorithms, such as (1) social bias, (2) user bias, and (3) informational bias.

(1) *Social bias in algorithms.* Algorithms can exhibit *social* biases and prejudices against people or stakeholders, whether individuals or groups, who are explicitly represented in an information system, or are directly affected by decisions reached using these systems (e.g., Hajian et al. 2016; O'Neil 2016). In cases of biased representation, some of the categories and inference rules for representing and reasoning about individuals and groups in the system contain social biases against them. For instance, an information system that determines solvency or creditworthiness of individuals on such factors as the creditworthiness of their neighbours and family members, or even on their ethnicity or immigrant status, is likely to replicate social prejudices in decisions about loans and ignore the unique features and circumstances of these persons. In cases of indirect social bias, stakeholders are not directly represented by the system, but there are nevertheless outcomes that tend to negatively affect certain individuals or groups more than others. For example, algorithms designed for the purpose of managing the social effects of natural disasters in cities may only represent the effects of disasters within the city's parameter and not the effects in the immediate vicinity of the city, thereby showing a bias towards aid for victims within the city's limits, leaving out potential victims outside the city's margins.

The unfairness that results from such systems can be analysed as unjust exercises of episodic power by actors who use the system to make decisions over persons and groups. It can be regarded as an unfair and biased exercise of asymmetric episodic power because it involves actors with the power to affect people's lives in a positive or negative way (e.g., by deciding whether they receive a loan, or whether they are in rescue operations), who decide and act on the basis of information provided by an information system which contains social biases, leading these actors to decide and act in a way that sustains these biases and leads to an unfair treatment of some persons or groups. Note that these actors need not at all be aware of these social biases. As Friedman and Nissenbaum (1996) have argued, social biases can enter into information systems in various ways, through conscious or unconscious prejudices of designers, through technical constraints, or through unanticipated contexts in which the system is used, and their users need not be aware of them.

(2) *User bias in algorithms.* User bias is bias in the usability of a system for different types of users (individuals, groups and organisations) (Brey 1998). By 'usability' Brey means the accessibility of the system for different types of users, the comfort and ease by which they can use its features, and the extent to which the system can satisfy their needs. User bias is different from the previously discussed type of social bias in that it does not pertain to stakeholders that are represented in the system, or that are affected by decisions and actions made by users of the system, but instead pertains to users of a system. Algorithmic systems can serve important needs for users, and if they systematically provide more benefits to some categories of users than to others, or even exclude some users from using the system altogether, this may in some circumstances be called unfair. Designers and developers have implicit ideas about the kinds of users for whom they are developing, what their values and interests are, and for what purposes and in which contexts they use the system. These presuppositions often exclude or disadvantage groups of users that do not fit this profile. But users may also be disadvantaged for arbitrary reasons, such as limitations of the technology in question. For instance, an algorithm that helps people find potential partners in an online database may not contain social biases in how it represents people, but it may be structured such that users with particular needs or interests (e.g., those for whom it is very important that a potential mate has the same religious background or level of education) are not well accommodated by the system as there are no algorithms to accommodate for these kinds of searches. Alternatively, the system may allow for searches and displays of personal information that violate privacy or confidentiality laws in certain countries, making the system less usable or not usable legally in these countries, or it may display information about persons that is considered taboo in particular cultures, thereby also limiting usability.

The unfairness resulting from user bias can be analysed as algorithmic systems engendering unequal distributions of dispositional power in users. If a system grants particular dispositional powers to some groups but not to others (e.g., the ability to find a partner online, or to navigate between points in space, or to do well in the stock market, etc.), then the resulting inequality in dispositional powers could be unfair. This is especially the case when what are at stake are dispositional powers that are important for pursuing one's goals and having a good life (equalling what philosopher John Rawls calls primary social goods) and when there are no equally good alternative means available to persons to acquire such powers.

(3) *Informational bias in algorithms.* Informational bias is bias in the selection and representation of information that is offered by a system (e.g., Brey 1998; Kraemer et al. 2011). Many of these biases are not social biases but rather cognitive biases in the way in which information is represented and selected that lead to distorted information about the world. For instance,

an algorithm designed to select and tag the most important objects in a photograph may disproportionally select objects with geometric forms. Or a system may be designed to simulate weather but not be able to simulate conditions of mist. Or an algorithm may be designed to calculate results from political polling but oversample certain groups. Other examples of informational bias involve cases where using different search engines can result in different and potentially biased search results (e.g., Vaughan & Thelwall 2004; Goldman 2006; Epstein & Robertson 2015; Nevelow Mart 2017; Noble 2018).

Informational bias can have ethical consequences, insofar as they can cause harm to users or other stakeholders. We will restrict ourselves here to users. AI systems, and information systems generally, are intended to empower their users by providing them with information that allows them to make better and more informed decisions and take better actions that further their interests. However, systems with informational bias and error may not empower their users and may instead end up disempowering them. Such disempowerment can lead to harms, including serious harms, that have ethical import.

6.3.2. Origins of Algorithmic Bias

Algorithms can also be distinguished on the basis of their origin, that is, the manner in which algorithmic bias has come about. There are three main ways in which algorithms can become biased: (1) by nature, (2) by nurture, and (3) by consequence.

(1) *Biased by nature.* This class of biased algorithms consist of cases in which algorithmic code and instructions are designed on the basis of prejudiced premises. Such a possibility can be observed when, for example, the prejudiced mindset of coders and software engineers finds its way into the design of algorithms by becoming embedded in the process of specifying the conditional and consequent parts of algorithms. Consider the following. Generally speaking, algorithms and algorithmic instructions usually come in the form of conditional statements, such as: IF (condition X) THEN (consequent Y). Specifying the conditions and the consequents of such statements gives rise to normative (e.g., if X then ought to do Y) and/or imperative (e.g., if X then do Y) types of sentences or statements. Insofar as ideology (whether religious or secular) and ideological power involves the construction and communication of such normative and imperative statements and narratives (e.g., Ricoeur 1986; Mann 1986), software algorithms can come to embody various normative or imperative statements and narratives.

Consider, for instance, UrbanSim, a software platform used for simulating and predicting urban growth (Waddell et al. 2003). Earlier—theoretical—versions of such urban growth models and simulators rested on specific assumptions about economic behaviour (e.g., agents possessing perfect market information or markets being perfectly competitive). However, because of observed market imperfections, the developers of UrbanSim adopted less stringent assumptions about economic behaviour in order to raise the predictive power of their software platform (Boyce & Williams 2015, 385–396). What is noteworthy here is that assumptions about markets being perfectly competitive and agents possessing perfect knowledge are indeed implicit within mainstream neoclassical economics, whereas such assumptions are considered to be value-laden and ideological within, most notably, institutional economics. Hence, what this example shows is that those earlier versions of urban growth simulators codified value-laden or ideological conceptions of market behaviour into algorithmic instructions. Once an ideologically laden statement or narrative becomes codified in the form of software algorithms as in the above example, they can become a means of perpetuating those ideological narratives.

Another example of algorithms biased by design is when using search algorithms designed by different programmers can result in different search results. For instance, in a recent paper, Susan Nevelow Mart (2017) considers algorithms as human artefacts. The author has compared the results of using the algorithms of the scholarly search engines Google Scholar, Westlaw, Lexis Advance, Fastcase, Ravel, and Casetext and has argued that "six groups of humans created six different algorithms, and the [variation in] results are a testament to the variability of human problem solving" (2017, 387). This variability across different search algorithms, according to Nevelow Mart, has significant implications for conducting research. Insofar as online scholarly research is concerned, this study attests to the imperative of not relying solely on electronic search and combining electronic and paper-based forms of research (e.g., Bintliff 2005, 23–25).

(2) *Biased by nurture.* This class of biased algorithms consists of cases in which algorithms (particularly machine-learning algorithms) happen to have been programmed to learn and reproduce prejudiced behaviour in their course of interactions with their social or informational environment. An example of algorithms prejudiced by nurture is the case of a chatbot called Tay, developed by the Microsoft Corporation, which operates on the basis of machine-learning algorithms and is thus designed to have chats with humans by mimicking human speech through reproduction of speech patterns of humans aged between 18 and 24 (Hern 2016). The chatbot in question was launched on the 23 March 2016 but was shut down after less than a day, because it began to tweet racist and misogynist comments (Hern 2016). Soon after switching off the chatbot, Microsoft issued a statement, saying that "the AI chatbot Tay is a machine-learning project. . . . As it learns, some of its responses are inappropriate and indicative of the types of interactions some people are having with it" (quoted in Hern 2016).

As an example of algorithms biased by nurture, we can also consider machine-learning algorithms used to assist human resources personnel in making hiring decisions and in mapping out vast pools of potential candidates. Such data-driven algorithmic software can be efficient in evaluating resumes, and, in some cases, can filter out up to 72 percent of job applications before a human resources officer can see them (Mann & O'Neil 2016). When such algorithms are designed to learn from human recruiters who happen to be prejudiced, these algorithms recreate existing prejudice. In a recent experiment, recruiters processed identical applications but selected more candidates with white-sounding names, such as Emily and Greg, than with non-white-sounding names, such as Lakisha and Jamal (Bertrand & Mullainathan 2004). Thus, when algorithmic hiring systems are nurtured on the basis of such already prejudiced data, they will make their decisions on the basis of subjective criteria and thus disregard the true potential of rejected candidates.

(3) *Biased by consequences*. This class of biased algorithms consists of cases in which algorithms turn out to be biased in their consequences (albeit the algorithms are not intended to be biased). Consider, for instance, a piece of algorithmic code that tells the software programme to select only highly educated candidates in the process of hiring for employment. But then it turns out that the algorithms favour, for instance, white males in a particular institutional context in which the pool of highly educated candidates happen to be white males. In this particular case, the algorithms are certainly biased in favour of highly educated people, but the algorithms are not programmed in this way; rather, they happen to be prejudiced in favour of "white males" in their consequences, which in turn depend on the particular institutional context in which such algorithms are embedded. Such cases of biased algorithms further demonstrate the importance of the institutional context in determining the prejudiced power of algorithms. Without the institutional context, the algorithms in question would not have the prejudiced consequences; it is only through their workings in a particular context that they become "prejudiced"—a context which is in turn shaped by existing norms, practices, and institutions (e.g., practices that favour and promote education for white males).

The foregoing discussion has aimed to show, first, how various biases can be programmed in the premises of algorithms; second, how algorithms can become prejudiced by acquiring and reproducing prejudiced behaviour through interactions with their social environment; and third, how algorithms can become prejudiced in their consequences depending on the particular institutional context in which they are embedded. Here, it should be emphasised that these three categories of prejudiced algorithms are analytic categories, since in reality they greatly overlap. While few algorithms are prejudiced explicitly in the code, prejudice becomes engendered in their consequences, when algorithms become embedded in and entangled with existing institutions, which has to do with the systemic and institutional character of power. Indeed, the bias and power of algorithms become further reinforced and their effects further exacerbated when algorithms become entangled with certain other already existing norms, practices, and institutions. Specifically, such a possibility is observed when, for example, technology users are subjected to an asymmetrical exercise of power by making them agree to the use of algorithms of dubious nature. Algorithms are frequently designed and produced by private firms and companies, with ordinary citizens and users having little say in their design. Although people are asked for their consent to the terms of service, they nevertheless result in coercive and asymmetrical relationships of power: one 'must' agree to the terms of use if one wants to use them.

6.3.3. Bridging Bias and Power in Algorithms

As noted in Section 6.1, the nature of the relationship between algorithmic power and algorithmic bias has not been explicitly addressed in the current social and philosophical literature on algorithms. This subsection undertakes the task of elucidating precisely how bias can relate to power in the context of increasing reliance on algorithmic data-processing, autonomous and semi-autonomous decision-making. It is proposed that we should analyse the relation between bias and power through the concept of exclusion. Briefly stated, the connection between bias and power becomes clearer when we consider the cases in which a social group becomes subjected to power through becoming excluded from the distribution of socio-economic resources and goods through mobilisation of bias against this group. The discussion below thus focuses on cases of algorithmic bias that result in the exclusion of certain individuals and groups from the distribution of resources, benefits, or other social goods.

The notion of the mobilisation of bias was introduced by Eric Schattsneider (1960) as a concept to understand the imbalances in the distribution of power in society. People in democratic societies are usually said to have access to power holders, whether elected representatives or unelected bureaucrats, by means of various interest- and pressure-groups that can lobby the politicians on people's behalf. In his book *The Semi-Sovereign People* (1960), Schattsneider argued that such access to politicians can be seriously imbalanced because—in his famous and often-cited words—"*organization is the mobilization of bias*. Some issues are organized into politics while others are organized out" (1960, 71, emphasis in the original). The main point here is that for a social group to have its issues heard or its interests represented, it must be sufficiently well organised to do so. Underlying the idea of the mobilisation of bias are the notions of exclusion and (its opposite) inclusion.

The 'organisation of issues into politics' entails the inclusion of the interests of certain groups in the political agenda, as opposed to the 'organisation of issues out of politics' which involves the exclusion of the interests of certain (usually marginalised, or less organised) groups from the political agenda.

The idea of the mobilisation of bias also applies to the cases of social and informational bias in algorithms discussed earlier in this chapter (see Subsection 6.3.1). For instance, algorithmic data-mining and data-processing systems, which are increasingly deployed within government and policy decision-making processes, are expected to help policy makers and officials to make better and informed decisions regarding a wide array of social and political issues. However, as noted earlier, such systems can be prone to informational bias and error, and using systems with informational bias can lead to misleading results and decisions. An example of this would be an algorithm that gathers and processes data about the extent of damage caused to civilians and their property by a severe weather condition (such as a hurricane) and fails to process the data relating to the damage caused to civilians residing beyond the city limits, thereby excluding this group of residents from the relief aid. In this regard, informational bias exhibited by algorithms would perpetuate the bias against, or in favour of, the issues and interests of one social group as opposed to some other social group in the course of government decision and policy-making processes.

Bias and its mobilisation thus make it possible to exclude certain individuals and groups from the distribution of political power, economic resources, and other social goods. On this view, bias relates to power through social exclusion. Therefore, to shed further light on the relation between bias and power, we should consider the notion of social exclusion in more detail. As a concept, social exclusion seems to have come into prominence in policy and academic literature in the context of discourses concerning social welfare politics in the 1980s, which, at the time, was characterised by increasing unemployment, straining of welfare funds, and calls for social justice (e.g., Levitas 2005; Bauman 2005; Ryan 2007). In that context, social exclusion of individuals and groups, according to Levitas (2005), has been accounted for within three different discourses: (1) a redistributionist discourse which draws attention to the poverty of the excluded; (2) a moral underclass discourse which focuses on the moral and behavioural character of the excluded; and (3) a social integrationist discourse which emphasises paid work as a means of integration of the excluded. Kevin Ryan (2007; 2011, 224–228) offers summarised interpretations of the three discourses as follows:

(1) *The redistributionist discourse.* The main thesis is that the (global) economic and political institutions by their nature consistently serve the interests of the dominant classes and elite groups. The excluded are reduced to a 'new

servant class' or marginalised social group, such as ethnic minorities, migrant communities, single mothers, and so forth, who have come to depend on low-paid manual service and sweatshop labour. On this discourse, according to Ryan (2011), impoverished conditions characteristic of underdeveloped countries and societies are now found in the public housing districts, inner city ghettos, or urban slums of the developed world. The main causes of social exclusion, according to this discourse, are capitalist institutions, neoliberal reforms, and government spending cuts and austerity measures, as well as the increasingly provisional and conditional character of welfare aid and entitlements.

(2) *The moral underclass discourse.* The main thesis is that the socially excluded are an underclass characterised by moral decay, behavioural delinquency, lack of discipline and responsibility. On this discourse, the excluded are not only those who happen to have become disconnected from the rest of society, but also those who actively oppose integration, social norms, or the rule of law; that is, exclusion becomes self-exclusion. In this way, assuming a moral decline, the discourse shifts the responsibility for the social exclusion away from economic and political institutions toward the excluded themselves. Moreover, socio-economic policies and programmes aimed at the eradication of poverty are considered to be counter-productive insofar as such policy efforts and structures are deemed to be perpetuating the conditions of socio-economic dependence and to be conducive to criminality.

(3) *The social integrationist discourse.* The main thesis is that the socially excluded are the accidental or unintended victims of globalisation, modernisation, or technological progress. The social integrationist discourse, according to Ryan, is a "sociological thesis of unintended effects" and "a discourse of renewal and reform" (2011, 225). On this discourse, the new age of economic and technological progress and development can benefit everyone if it is acknowledged that everyone has both rights and responsibilities. Individual empowerment is regarded as a way to social empowerment. On this view, social integration can be achieved by devolving responsibility to individuals, and by transforming them into consumers of private and public services. Yet, as Levitas (2005, 26–27) notes, this discourse obscures social, class, and gender inequalities by overemphasis on integration through paid work.

The above discourses summarise the dominant ways in which the problems of social exclusion have been evaluated and accounted for within academic and popular discourses in the recent decades. It is therefore possible to use these discourses of social exclusion as a starting point for mapping the possible directions which might be taken by the emerging discourses on algorithmic bias as a factor in the perpetuation of existing social divisions in society. Below, in a tentative and provisional manner, we shall modify the three discourses of social exclusion to the case of algorithmic bias and exclusion. Namely, the redistributionist discourse will be recast as the 'exclusion-by-design' discourse; the moral underclass discourse as the 'exclusion-by-character' discourse; and the social integrationist discourse as the 'exclusion-by-accident' discourse. The resulting discourses can be outlined as follows:

(1) The 'excluded-by-design' discourse. The core argument here would be that the organisations and institutions responsible for technological development, and other sites of technological innovation are deeply embedded in the broader capitalist, neoliberal, and patriarchal power structures which create and reproduce an underclass of individuals and groups marginalised on the basis of their race, gender, ethnicity, or disability. Consequently, algorithms developed and implemented in the context of such institutions and sites of innovation would tend to be biased against this underclass consisting of migrant communities, ethnic minorities, single mothers, the disabled and the infirm. On this discourse, responsibility for the bias and exclusion engendered by algorithmic technologies and practices would be firmly ascribed to the dominant economic and political institutions, corporations, and policy makers.

(2) The 'excluded-by-choice' discourse. The core argument here would be that the problem is to be found in the inability or unwillingness of the excluded and affected individuals and groups to adapt to the new algorithmic realities. The excluded-by-choice discourse would tend not to acknowledge the ethical issues of algorithmic bias, by shifting the focus on the excluded people 'who have chosen not to conform'. Consequently, the burden of responsibility would be placed on the excluded individuals and groups, instead of designers, corporations, government agencies, and other institutions. Notably, policy measures aimed at offsetting the loss of opportunity suffered by the excluded groups as a result of algorithmic bias would be regarded as creating the conditions that further perpetuates their criminality, delinquency, or social and economic dependence.

(3) The 'excluded-by-accident' discourse. The core argument would be that the ethical issues associated with algorithmic bias are the by-product of still evolving algorithmic technology, unforeseen or unpredicted by its designers, policy makers, or whoever was involved in its development and implementation. The excluded are merely the victims of (possibly inevitable) progress of AI and ML algorithmic technologies. Within this discourse, the issues of algorithmic bias can be viewed as something that can be resolved in future, for example, through further reconfiguration or improvement of the algorithmic technology. At the same time, this discourse would emphasise the importance of adapting to the new algorithmic realities by the excluded (e.g., by learning or acquiring new skills and qualifications) as a means of re-integration with the rest of society.

The discourses sketched above are formulated not only as a prediction of how discussions of algorithms can play out in future discourses on algorithmic bias and exclusion. In varying forms, they can already be found within current discourses on technology in general and algorithms in particular. The three discourses discussed above articulate more or less specific strategies and programmes concerning the design, uses, and effects of algorithmic technologies and practices. As such, these discourses constitute, and are constituted by, a domain of social and political struggle over the meaning and interpretation of the various relevant aspects of algorithms in society. Although here we could only offer outlines of these discourses (for obvious reasons of space), these sketches can serve as a point of departure for further theoretical and empirical explorations of the ethics and politics of algorithmic technologies and practices. Importantly, the discussion has highlighted the possibility that the design, uses, and consequences of algorithms are, at least to some extent, dependent on how different discourses play out within the field of political struggle. In this regard, the position sketched above has some affinities with the social constructivist approach of Donald MacKenzie in his recent study of algorithms in the context of high frequency trading (2018).

6.4. CONCLUSION

To sum up, the notion of algorithmic power is fast becoming an object of both academic and policy discourses. Within these discourses of particular concern is the issue of so-called prejudiced algorithms. Moreover, within such discourses one can observe a number of different senses of algorithmic power being employed, especially when viewed from a multi-disciplinary angle. These developments showcase an urgent need for the creation of a philosophical framework for clarifying and understanding different conceptions of algorithmic power. In this respect, the present chapter has applied the philosophical framework developed in the first part of the book to the case of algorithms with special attention to the issues of algorithmic bias. This allows for a differentiated vocabulary for describing various senses of algorithmic power, while bridging together critical studies of algorithms, socio-political theory and philosophy of action and technology. Any adequate ethics of technology, practiced here as an ethics of algorithms, must be based on a better, clearer, and more nuanced and differentiated understanding of the many ways in which they can be described as 'powerful' and 'prejudiced'. The discussion furthermore suggests that a more unified account of ethical issues in algorithms can be given through a pluralist theory of power. Various ethical issues can be identified as resulting from changes and realignment of episodic, dispositional, systemic, and constitutive powers in users, represented persons and groups, and other stakeholders. These changes in empowerment and power relations can cause harm, violate rights, and lead to inequalities and injustices. For designers, it may be easier to deal with ethical issues in the design of algorithms if they can be translated into aspects of power. Designers can study how algorithms empower or disempower stakeholders by granting them enhanced abilities or by taking them away, and how they may result in asymmetric power relations by granting users more control over persons or groups. In the same way, users, policy makers, and others can study empowerment and power relations to draw similar conclusions and take appropriate action. While an account of ethical issues in algorithms cannot fully replace ethical assessment, it can help such ethical assessment by making clear the relations between users, algorithms, and institutions.

Note

1. The possible future implications of the regulation are still being debated (e.g., Wachter et al. 2018). Nonetheless, the point of this example is that legal and judicial regulation and its effective implementation within the bounds of a national jurisdictions can have significant effects on emerging algorithmic technologies and practices.

Chapter Seven

Power and Ethics

Conceptual Connections

This chapter continues exploring the role of the concept of power in the ethics of technology. While the previous chapter has examined the role of the concept of power in the nascent field of the ethics of algorithms, the present chapter considers how power relates to a set of (mainly ethical) concepts that have been prominent in recent ethical studies of technology. These concepts include responsibility, vulnerability, authenticity, and trust. The following discussion makes the case that there is a close conceptual connection between power and these concepts, at least insofar as these concepts somehow assume or imply presence or absence of power in individuals and groups to whom they are applied. For example, ascription of responsibility often implies that the responsible person or group had some power to bring about a certain outcome or has power to bring about some future outcome. Similarly, attribution of vulnerability frequently implies that the vulnerable person has no power, or that she is vulnerable in relation to some other powerful person. As these examples suggest, power can be regarded as a primitive, or foundational, concept which is presupposed by other less primitive concepts, such as the ones mentioned. The chapter thus explores the conceptual connections between power and these concepts.

7.1. POWER AND RESPONSIBILITY

Let us begin by considering the concept of responsibility. The concepts of power and responsibility can relate in a number of ways. For example, following the famous moral dictum that 'ought' implies 'can', it is held that

117

one's moral responsibilities must not exceed one's power to perform or satisfy those responsibilities. Similarly, when an agent is attributed responsibility for bringing about a certain moral outcome, it is also assumed that the agent had sufficient power to bring about the outcome in question. (In such cases, lack of power can serve as a basis for rejecting responsibility, as we often see in claims involving alibis: She could not have done it even if she wanted to, because she lacked power to do it.) The latter is a case of so called 'backward-looking' responsibility. There are also cases of 'forward-looking' responsibility. For example, according to the famous dictum that 'with (great) power comes (great) responsibility', the powerful can have responsibility to remedy some morally or socially bad or reprehensible situation, although this powerful actor might not be the one who brought about the situation in question. Let us consider these connections in more detail.

William Connolly is one of the authors who provides an interesting analysis of the relationship between power and responsibility. In *The Terms of Political Discourse* (1993), Connolly offered an essentially episodic conception of power (1993, 88), or what he himself called a "paradigm" form of power, according to which "A exercises power over B when he is responsible for some x that increases the costs, risks, or difficulties to B in promoting B's desires or in recognizing or promoting B's interests or obligations" (1993, 102–103). Central to his episodic conception of power is the notion of responsibility; or, as Connolly himself put it, "[T]here is an intimate connection between alleging that A has power over B and concluding that A is properly held responsible to some degree for B's conduct or situation" (1993, 95). Connolly supports this conclusion by arguing that the distinctions we draw between different forms of episodic power, such as coercion, manipulation, and persuasion, are "built around, and reflect, shared moral considerations", whereby "the distance between persuasion and manipulation is a moral distance; it reflects the judgment that there is a moral presumption against the latter that does not obtain for the former" (1993, 94). The main point here is that when A has exercised power over B, A can be held responsible for B's subsequent behaviour or conduct to the extent to which A has limited B's freedom or ability to choose or to act. On this formulation, A bears more responsibility for B's conduct when A coerces B than when A persuades B, insofar as the limit imposed on B's capacity for choice is far greater in the case of coercion than in the case of persuasion. Thus, much of the point of the concept of power, for Connolly, lies in its use for assigning responsibility. The concept of power, as argued by Connolly, is less suited for predicting the future behaviour and choice of social actors and groups. Instead, for Connolly, power offers "a perspective that enables [us] to locate responsibility for the imposition of limiting conditions by linking those conditions to the decisions people make, or could make and don't" (1993, 101).

This interpretation of the connection between power and responsibility is more applicable to the episodic conception of power, the conception of power that defines power in terms of its exercise. On this view of power, responsibility is linked to the actual exercise of power. A blackmailer who has not yet coerced her victims into surrendering their expensive belongings cannot yet be held responsible for a (future) coercion which has not yet taken place and which may not even take place. Only when the blackmailer has actually coerced her victims to do what the blackmailer wants, then we can, other things being equal, assign responsibility to the blackmailer. This example suggests that Connolly's interpretation of the connection between power and responsibility would be untenable in terms of the dispositional view of power which defines power as abilities and potentialities.

Peter Morriss, a strong advocate of the dispositional view of power, has indeed challenged Connolly's analysis of power and responsibility, by arguing that "if A could, *but does not*, 'limit' B in some way, then there is nothing in B's conduct or situation for which A can be held responsible: ex hypothesi, A has not contributed to the situation that B faces. It is only when A does act that he can be held responsible for the act's consequences" (2002, 21; emphasis in the original). Since Morriss views power as referring to dispositions and not actions (or its consequences), he concludes that Connolly failed "to connect responsibility with any concept recognizable as power" (Morriss 2002, 38). Thus, for Morriss, who defines power strictly as an intentional disposition to effect relevant outcomes, the business of holding people responsible "invariably" involves their already performed actions, not their powers.

Having thus argued, Morriss formulated an alternative view of the connection between responsibility and power, formulated in dispositional terms. According to this alternative view, there are two ways in which responsibility can be connected with power as a disposition. The first connection between power and responsibility, for Morriss, concerns denial of responsibility for some action or outcome. We are not usually responsible for something we did not do. For example, if a post office gets robbed somewhere in London, we can deny all responsibility by demonstrating that we were in the Shetland Islands at the time of the robbery, and that robbing post offices over such great distances is beyond our powers. As Morriss notes, "[A]ll claims involving alibis contain such an implicit reference to lack of power" (2002, 38).

The second connection between responsibility and power, of which Morriss writes, finds its expression in the following statement made by Lord Salisbury: "Those who have the absolute power of preventing lamentable events and, knowing what is taking place, refuse to exercise that power, are responsible for what happens" (quoted in Eccles 1981, 389). Morriss thus argues that one can be held morally responsible for some negative outcome that one could have prevented but did not prevent it knowingly. Note that in this latter case, it is not *having* power but *failing to exercise* that power that makes one responsible, for "there is nothing whatsoever wrong with having the power to prevent lamentable events—indeed, the more people there are with this power the better. What is wrong is having the power and refusing to use it" (Morriss 2002, 39). Considering the above two cases, Morriss concludes that the relationship between responsibility and dispositional power is of negative character: "[Y]ou can deny all responsibility by demonstrating lack of power. You can do this, as an alibi example, by proving that you couldn't have *done* the crime. Or you can do this by showing that you couldn't have *prevented* the catastrophe. In either case, power is a necessary (but not sufficient) condition for blame: if you didn't have the power, you are blameless" (2002, 39).

Connolly and Morriss's analyses of the connection between power and responsibility does not exhaust the space of all possible relations between the two concepts. Specifically, both Connolly and Morriss seem to be interested in what is called 'backward-looking' conception of responsibility, which concerns the apportioning of blame, liability, or accountability to someone for a certain action performed or an outcome produced. Backward-looking notions of responsibility are contrasted with forward-looking notions of responsibility, which can be understood in terms of an obligation or duty to ensure that a certain action or outcome takes place. Traditionally, much of philosophical discussions around responsibility has attended to backward-looking notions of responsibility (Van de Poel 2011). Recently, however, some authors began focusing on forward-looking notions of responsibility, for example, from the perspective of utilitarianism (Goodin 1995) or virtue ethics (Ladd 1991; Williams 2008; Van de Poel 2011). To get a better grasp of the distinction between backward-looking and forward-looking types of responsibility, it is helpful to consider the following case-study offered by Nihlén Fahlquist (2017):

> In 1997, the Swedish government adopted a new goal to guide traffic safety policy. It was called Vision Zero and the basic idea was that it is not ethically acceptable to have numerical goal stating how many annual fatalities are acceptable in road traffic. . . .Instead, it was argued that the only ethically justified goal is that no one is killed in road traffic. . . .The new framework acknowledges that a driver may be said to have causal responsibility for a collision, but the accident additionally entails a forward-looking responsibility for someone else, in this case the so-called system designers, that is, for example, local government and car producers. An individual driver may have caused a collision through her inability to notice a pedestrian at a crossing. This inability could potentially be reduced, for example, by vehicle technology, road design, or street lighting. (Nihlén Fahlquist 2017, 134)

In this particular case, as Nihlén Fahlquist argues, the system designers and government institutions are not the ones who caused the accident, but they have a responsibility to ensure that the traffic is designed in such a way so as to prevent similar accidents in the future. This case illustrates a shift from the backward-looking responsibility of the individual road users to the future-looking responsibility of the system designers (Nihlén Fahlquist 2006). The case is also illustrative of the need to make another distinction, that is, the one between individual and collective (or, institutional, systemic) responsibility. While individual actors may share in the responsibility for bringing about certain outcomes, collective entities, such as institutions, communities and governments, may be better positioned, or have effective power, to ensure that such outcomes do not take place in the future. The distinction between individual and institutional responsibility is particularly important when we consider the socio-technically distributed character of agency and power within highly complex and ever-changing environments (Simon 2015).

As Hans Jonas (1984) contended in *The Imperative of Responsibility* (1984), traditional notions of morality, as found in both ethical theory and 'folk morality' (i.e., the moral principles of ordinary individuals), fall short in morally evaluating many of the actions related to the development and employment of modern technology, because they presuppose a notion of action that does not fit the type of action found in dealings with modern technology. These traditional modes of conceiving responsibility have thus become less suited in the moral appraisal of modern human societies increasingly characterised by processes of technological, economic, and political inter-connectedness and globalisation. In this context, we need new ways of thinking about responsibility, and shifting our philosophical focus onto forward-looking notions of responsibility seems to be a step in the right direction. Importantly, a good starting point for intellectual and practical considerations of such responsibility concepts can be the question of who, or what entity, under what conditions is best positioned to have the ability to do something about the immense technological, ecological and other issues we face today, which is a question of power.

7.2. POWER AND VULNERABILITY

In the literature on power, there has been a tradition of distinguishing between active and passive forms of power. John Locke, for example, wrote, "Fire has a power to melt gold . . . and gold has a power to be melted. . . . Power thus considered is twofold, viz. as able to make, or able to receive any change. The one may be called active, and the other passive power" (Locke 1854, 359–360). Similarly, Thomas Hobbes distinguished between an *agent* and a *patient* of power, where the agent is a body that does something to another body, that is, the body of the patient. According to this distinction, active power (i.e., the power of the agent to produce change) and passive power (i.e., the power of the patient to undergo change) are necessary counterparts to each other. The existence of one necessarily requires the existence of the other, so to speak.

The distinction between active and passive powers might not stand up to metaphysical scrutiny, since it is, as Peter Morriss notes, "unacceptably animist" (see Morriss 2002, 99–102) for a different interpretation of the distinction). However, a distinction along these lines can make some sense when it is applied to episodic power relationships, such as seduction or coercion (as discussed in Chapter 2), where 'passive powers' can be interpreted as vulnerabilities. In episodic power relationships, a power holder is an 'agent' exercising 'active power' to coerce or seduce, whereas a power endurer is a 'patient' possessing 'passive power' to be coerced or seduced. Such passive powers here can be interpreted as qualities that render their possessors vulnerable to the acts of coercion or seduction. For example, in a coercive relationship, A's threats of punishment to B are effective insofar as B finds herself vulnerable to these threats. Similarly, in a seductive relationship, A's promises of reward to B are effective as long as B finds herself vulnerable to these rewards. Thus, a passive power is a vulnerability.

This connection between power and vulnerability can become clearer when we consider the so-called 'zero-sum' character of episodic power relationships. As noted in Chapter 2, episodic power relationships are often described as being 'zero-sum' (or sometimes 'negative-sum'), meaning that within episodic power relationships an increase in the power of the power-wielder corresponds to a decrease in the power of the power-endurer (e.g., Castells 2009). For example, if my power to coerce you increases, it means that your power to resist my coercive advances become decreased; or vice versa, if your power to resist my coercive actions increases, then my power to coerce you decreases. On this zero-sum view of episodic power relationships, a decrease in one's power to resist others can be understood as a vulnerability. Within such relationships, a power holder increases her power by exploiting the vulnerability of the power-endurer; or vice-versa, the power of the power holder is decreased through elimination of the vulnerabilities of the person-endurer.

The presence of power and vulnerability can be observed in relationships between custodians and dependents, doctors and patients, researchers and research subjects, and so on. This is how, according to Ryan Calo (2016), the law often conceives of the vulnerability (i.e., as a product of some special relationship). In such relationships, the power of the custodian to abuse the dependent increases as the power of the dependent to resist the abuse decreases, which also means an increase in vulnerability on the part of the dependent. Such relationships therefore frequently require certain institutional measures in order to limit the potential abuse of power on the part of the powerful with the aim of protecting the vulnerable. For example, in the context of privacy law, Jack Balkin (2016) discusses how custodians of sensitive consumer information should be considered "information fiduciaries" who should have obligations of loyalty and care (also cited in Calo 2016). Here, an attribution of vulnerability in one actor implies the possession of power by some other actor, and vice versa.

In the above examples, the relation between power and vulnerability are cast in terms of the episodic view of power which regards vulnerability as either present or absent depending on the (exploitative) actions of the powerful. However, this view of power and vulnerability tends to limit our attention to the actual (or very imminent) exploitation of vulnerability. Consequently, this understanding would tend to endorse the view that vulnerability merits legal or policy attention only insofar as it requires protection from exploitation, whereby the actual eradication or elimination of vulnerability would tend to fall beyond the scope of law and policy. For this reason, perhaps, we should conceive of vulnerability in dispositional terms; namely as a dispositional feature, property, or characteristic that people might possess irrespective of whether they find themselves in some potentially coercive, manipulative, exploitative, or abusive power relationship. On this understanding, vulnerability is something that we all have, whether young or old, able-bodied or disabled. One can be powerful yet be vulnerable. (Indeed, history is replete with cases in which even the most powerful politicians succumbed to political blackmail, *kompromat*.) Vulnerability is an inevitable constituent of the human condition. As Mark Coeckelbergh (2013) notes, human struggle against this feature of the human condition is neither an exclusive feature of the modern scientific and technological project to make life artificial, as Hannah Arendt wrote in *The Human Condition* (1958), nor it is of the 'cybernetics' movement, as Jean-Pierre Dupuy argued in *On the Origins of Cognitive Science* (2009). Vulnerability is "part of what we are and what we do as humans" (Coeckelbergh 2013, 2).

Conceiving of vulnerability in dispositional terms sheds further light on the latent character of vulnerabilities. In Chapter 3, following Morriss (2002, 57–59), we have defined latent dispositional powers as abilities that can be obtained over time. For example, ability to speak a foreign language is a latent ability which can be obtained over time. Just like dispositional powers can be latent, vulnerabilities too can be characterised by their latency. While some of our vulnerabilities are known to us, there can be latent vulnerabilities which take time to develop. To illustrate the point, consider the latency of privacy vulnerability. According to Cory Doctorow, privacy is one of those areas of concern in which we frequently fail to connect causes and effects of privacy vulnerability, since acts of information disclosure and their consequences for privacy are often separated in time and space. As a case in point, Doctorow (2012) tells the following story:

> In the early 1980s, I had a teacher whose wife went into hospital to deliver their first child. Afterwards, they were approached by a nice man from a marketing consortium offering a basket of free nappies, baby-grows, wipes, and other necessities. All he wanted in return was the child's name, date of birth and address (details that privacy detractors trivialise as "tombstone" information). They gave it to him. A few weeks later, the baby died. It was unforeseen and tragic. More tragic, though, was what happened every year on the child's birthday: the grieving family got a slew of commercial offers in the post, targeted at a dead child's ageing ghost.

In this illustration, the effects of information disclosure took time to develop into a pernicious privacy issue. Many of those people who knowingly or unknowingly disclose their private information may not necessarily come to suffer any serious issues. However, the possibility of worst-case scenarios involving fraud, identity theft, or political violence should be sufficient to make people cautious about latent privacy issues. Yet, privacy is not the only area of concern characterised by such latent vulnerabilities. Public health, as Doctorow (2012) notes, is another such area where people often fail to connect the causes and effects when they smoke, drink, and overeat. If only the effects of such indulgences presented themselves soon enough, he muses.

Conceiving of vulnerabilities in dispositional terms also suggests a way of further expanding our understanding of vulnerability, when we consider the distinction between generic and time-specific abilities as dispositional powers. As shown in Chapter 3, generic abilities acquire their political significance when they are embedded into certain time- and place-specific circumstances. For example, someone with computer hacking abilities, but with no access to appropriate network or infrastructure, would not be able to steal and leak emails from the Democratic National Committee, with significant political consequences. On this view, generic abilities become time-specific abilities only within specific technological, institutional, and political settings. This same applies to vulnerabilities as dispositions. Consider dark skin colour as an example. It is a naturally occurring human skin colour characterised by the presence of eumelanin pigments in human skin. However, within certain cultural, political, or institutional settings, the skin colour becomes a source of serious vulnerabilities (e.g., Cooper 2015; Dowd 2013). This example suggests that humans are all vulnerable "in degrees and according to circumstance", with "some individuals and groups within society are more vulnerable than others" (Calo 2016).

7.3. POWER AND AUTHENTICITY

Authenticity, generally speaking, is an important and complex notion, which has a wide range of applications within diverse areas of ethics and moral philosophy. For example, the notion of authenticity plays a role within debates and discussions on human enhancement and enhancement technologies (e.g., Parens 2005; Bolt 2007; Juth 2011; Kraemer et al. 2011; Mackenzie & Walker 2015; Iftode 2019). It has also been argued that authenticity is a fundamental feature of human flourishing and well-being (e.g., Sumner 1996; Haybron 2008) and that authenticity mediates the positive effects of power on subjective well-being (e.g., Kifer et al. 2013). There are also more sophisticated accounts of authenticity developed by existentialist philosophers, who consider authenticity as necessary for understanding human existence, and conceive of authenticity as both an ideal of human flourishing and a normative theory of the self (e.g., Guignon 1993; Taylor 1991).

Importantly, the notion of authenticity often serves as a qualifier of values espoused by individuals and social groups, as observed in the distinction between authentic and inauthentic values, where authentic values refer to values which can properly be described as the agent's own values (Varga & Guignon 2017). As such, articulating a clear and coherent notion of authentic values is argued to be an integral element of analyses and accounts of personal autonomy (the latter being another crucial concept of ethics and moral philosophy (e.g., Christman 2018; Hyun 2001). Furthermore, the notion of authentic values also applies to the processes of making and enacting decisions about the design and implementation of technologies. For example, there is a class of value-centric methods and approaches to technology design—such as Value-Sensitive Design (VSD; Friedman et al. 2008), Disclosive Computer Ethics (DCE; Brey 2010), and so on—which aim to identify user and social values and inscribe them in the design of technologies. It can plausibly be argued that such approaches should be sensitive to the authenticity of values as they are embedded in the design of technologies, at least insofar as incorporating inauthentic values in the design of technologies would negatively impact the autonomy of their users.

Motivated by these concerns, our discussion here focuses on the connection between power and authenticity, by exploring how power can affect the authenticity of values, thereby having implications for personal autonomy and the values needed for guiding technology design. We can begin by noting that the general thesis that power can influence and shape social and moral values is not entirely novel. For example, Immanuel Kant is said to have maintained that "the possession of power unavoidably spoils the free use of reason", and in so doing affect moral values and judgement (quoted in Ash 1995, 39). Similarly, Lord Acton, in reference to the role of moral judgments in history, wrote that "power tends to corrupt and absolute power corrupts absolutely" (Lord Acton 1887). Yet, by far the most worked-out classical account of this thesis belongs to William Godwin, an eighteenth-century English philosopher and novelist. Godwin held a firm belief that powerful social and political institutions mould not only the actions of individuals, but also their values and opinions. Godwin specifically criticised previous thinkers for their having "treated morality and personal happiness as one science, and politics as a different one", and for their having "considered the virtues and pleasures of mankind as essentially independent of civil policy" (Godwin 1976/1793, 81).

In his *Enquiry Concerning Political Justice* (1976/1793), Godwin argued, in a sincere and uncompromising fashion, that powerful social and political institutions, frequently through their embodiment within different forms of government, have deleterious effects on the actions, attitudes, and moral values of people as citizens and subjects. The particular thesis that different systems of government produce different kinds of attitudes and value judgments in the minds of their subjects has been around for some time, and it can be found not only in Godwin but also in Plato, Aristotle, and Montesquieu. The latter of the authors, for example, distinguished three systems of government, each corresponding to its own distinctive "spirit": (1) despotism producing fear, (2) monarchy producing honour, and (3) republic producing virtue (Montesquieu 1989). Nonetheless, Godwin appears to have gone even further than these authors, regarding all three systems of government as corrupting the values of their subjects and citizens. Even monarchy, with its spirit of *honour*, was considered by Godwin as having baleful effects on individual values and judgment. In *Caleb Williams* (Godwin 2009 [1794]), one of Godwin's most accomplished fictional novels, the character Falkland represents a person of honour. Although Falkland seems to be a gifted, amiable, and benign person, he is prepared to commit crime and murder if necessary.

Political and social institutions of government, according to Godwin, can affect individual judgment and values in at least three ways (e.g., Monro 1953). Firstly, institutions create false and unnatural barriers among people, whether individuals or groups. Such barriers are usually the consequence of social and economic inequality, for example, engendered by the institution of private property. "There is nothing", Godwin wrote, "that more powerfully tends to distort our judgment and opinions than erroneous notions of concerning the goods of fortune" (Godwin 1976, 701). Accumulation of private property, according to Godwin, fosters in people "a servile and truckling spirit" (1976, 725), while unequal distribution of property results in a "continual spectacle of injustice" (1976, 727). Secondly, political institutions make people to do the right thing for the wrong reasons. For example, threats and acts of punishment, which make people comply with the law and the norm, have their powerful normative effects because people are coerced into compliance, and not because people appreciate the reasons or rationality underlying these norms. Therefore, what is wrong with coercion and punishment is that they eliminate the opportunities for people to acquire capacities to decide for themselves what is right and what is wrong and to guide their own behaviour in a rational manner. Finally, government institutions prod people to accept and submit to the opinions of the majority folk, or to the judgments of some minority rulers, without them properly apprehending the evidence or rationality upon which these opinions and judgments have been grounded. Hence people act from bias, prejudice, or prejudgment, as their views of other people and things become distorted as a consequence of generalisations, stereotypes, and the like. Particular categorisations of legal institutions, such as pauper, thief, or murderer come to serve as distorting glasses (1976, 261).

The three cases sketched above show government and its institutions as constraining the development and exercise of human capacity to live in accordance with one's private judgment. While there can be "something in the nature of man incompatible with absolute perfection" (1976, 145), universal benevolence, for Godwin, is an ideal worth striving for. One way to facilitate the exercise of private judgement by individuals is to limit the scope of interference of powerful political institutions. This conclusion finds its expression in Godwin's vision of the egalitarian society of the future, in which all institutional interference with private judgement is gradually eliminated. In this respect, Godwin can be argued to have shared some affinity with Kant, for whom political institutions had implications for moral autonomy. The development of individual capacities for self-regulation, as Kant saw it, required the elimination of constraints imposed by power that can foster desire against reason. Consequently, Kant is now known as an advocate of limited government (although what constitutes a limited form of government for Kant is not the same as the radically egalitarian view of society advocated by Godwin). In this way, both Godwin and Kant believed that powerful institutions can impose constraints on individual reason and judgement.

This (Godwin's) account of the deleterious effects of power on values also finds support in more recent accounts of authentic and inauthentic values. For example, Insoo Hyun (2001) has argued that what distinguishes authentic values from inauthentic ones is the circumstance in which such values have been acquired. What matters in distinguishing authentic from inauthentic values, for Hyun, is the absence or presence of specific "inimical, external influences" on these values (2001, 201). According to Hyun, "a person's value is authentic only if it is false that he is compelled to have it, and, by contraposition, that it is inauthentic in the case that he is compelled to have it" (2001, 203). Contrary to the critical reflection approach, which defines authentic values as those which have undergone the process of critical reflection by the individual, Hyun argues that it is better to define authentic values "negatively" as not having been acquired as a result of being compelled within restrictive social and institutional environments. Individuals, for Hyun, acquire inauthentic values when they are compelled to have them in the context of certain social institutions which deny these individuals unrestrained access to all other reasonable and relevant alternative values, views, beliefs, attitudes, or information (2001, 203).

To illustrate this point, Hyun discusses the cases of a happy slave and a happy farmer. Imagine a happy plantation slave living in the U.S. Deep South in the early nineteenth century, who values being an obedient servant of her master, not because she is coerced by threats of severe punishment, but because she believes that being obedient in this way is very important. In this instance, we have good reasons to question whether or not the happy slave is compelled to have these particular values. According to Hyun, it is certainly the case that the institution of slavery makes it very difficult, if not impossible, for the slave to pursue and develop other morally acceptable values and interests. All reasonable alternative values and interests, which are available to non-slaves, are out of the reach of the happy slave. We therefore conclude that her values are inauthentic, for she has no other alternatives.

The case of the happy slave stands in contrast to the case of a happy farmer. Suppose there is a happy farmer who is entirely satisfied with her way of life. She never questions her values for the simple reason that she never felt unhappy, but she can stand by her values if she is ever challenged. As Hyun notes, few would doubt the authenticity of her values. But what makes the farmer's values authentic and the slave's values inauthentic, given that both of them may be equally unreflective toward their values? The answer, Hyun argues, is that the farmer is free from the sort of pressures and influences which affect and restrain the slave. If the farmer ever becomes dissatisfied with her way of life, she can always move on to something else. As long as her values are in some harmony with her basic character and dispositions, there is little reason for us to think that she is compelled to have them.

There are a number of advantages of the account of authentic values sketched in the above paragraphs. First, it is an inclusive, non-elitist account that allows different people from diverse backgrounds, and not just philosophers and intellectuals, to have authentic values. Second, it helps explain why we make different evaluative judgements about authenticity of values, since inauthentic values assume a degree of coercion, compulsion, or constraint. Finally, it fits nicely with our everyday evaluations of certain core commendable values, such as parental concern for the well-being of children, which are not normally subjected to critical reflection. All in all, the main issue with inauthentic values is not so much whether individuals have critically reflected on them, but that in the process of acquiring inauthentic values these individuals have been wronged through coercion, compulsion, or constraint (Hyun 2001).

Having outlined two accounts of how powerful institutions can affect the authenticity of social and moral values, we can now consider the importance of authentic values to the processes of technology design. As noted earlier, approaches to technology design, such as Value-Sensitive Design or Disclosive Computer Ethics (see also Section 8.4), aim to identify and inscribe user values in the design of technology. In the context of such value-centric design approaches, the impact of power on values can have both moral and political implications. The moral dimension concerns the issue of authentic values and their importance for personal autonomy. If target users of a technology are compelled (in the words of Hyun) to have certain inauthentic values, then inscribing such values in the design of that technology would negatively impact the autonomy of these individuals, once they are implemented. The political dimension, on the other hand, concerns the potential of technology design to challenge or consolidate existing relations and structures of power (e.g., Simon 2017). Should inauthentic, or otherwise undesirable, values become embedded in the design of technologies, then such values would be 'recycled' through the process of technology design, instead of being resisted, with the whole process of design contributing to the perpetuation of institutions that are based on these undesirable values.

7.4. POWER AND TRUST

Imagine our contemporary world without trust. In the absence of trust, even the most mundane of things that we do in our daily lives, such as stepping into a cab, travelling by train, boarding a plane, paying a bill with a credit card, and so on, would be hardly possible. Without trust we would probably doubt that we know our own names. Trust can be viewed as something that binds people in their social relations, thereby holding society together (e.g., Luhmann 2017). In her discussion of trust, Sissela Bok noted that "whatever matters to human beings, trust is the atmosphere in which it thrives" (Bok 1979, 33). In the words of Annette Baier, "we inhabit a climate of trust as we inhabit an atmosphere and notice it as we notice air, only when it becomes scarce or polluted" (Baier 1986, 234). Despite the pervasiveness of trust in our daily lives, it is only very recently that trust began attracting attention in the philosophy and ethics of technology (e.g., Simon 2010; Nickel et al. 2010; Åm 2011; Nickel 2011, 2013, 2015; McCall & Baillie 2017; Sattarov & Nagel 2019).

Trust is often understood as a dispositional attitude or stance that others (whether humans, institutions, or technologies) will act or behave in a predictable manner that respects one's needs, values, and interests (e.g., Hardin

2002). On this view, a trusting attitude is formed on the basis of beliefs and expectations about the potential actions of others. These beliefs and expectations can be *predictive* about how others will act, and *normative* about how others should act (Nickel 2013). The trusting person's expectations about the trusted person's behaviour and actions originate from the former's assessment of the latter's interests and motives. Put differently, A trusts B with regard to some X, insofar as B's interests incorporate, or encapsulate, A's interests (Hardin 2002). This means that trust stems from the perception or belief that the interests of the trusted person respect those of the trusting person: if the values and interests of a person whom we trust do not conform to our values and interests, then our trust in that person cannot be rationally justified (Hardin 2002). Our expectations about how others will or should behave thus stem from our interpretation of other people's values and interests, which can be communicated through trustee's reputation, her present or past behaviour, as well as shared understanding of norms, roles, and responsibilities (Cook 2005).

The rational conception of trust sketched in the above paragraph tends to view trust as an intentional mental feature, attitude, or disposition that one actor has towards some other person or entity. However, it is possible to conceive of certain contexts, within which trust reveals itself as a property of a relationship between two or more people (see e.g., Niker & Specker Sullivan 2018). Conceiving of trust as a feature of certain kinds of relationships offers us a perspective for theorising the role of trust within, for example, asymmetrical power relationships (whether such relationships involve people, institutions, or technology). Certainly, given the complexity of the notions of power and trust, it is quite likely that there is more than one way to conceptualise the relation of trust and power. However, here we shall focus on trust vis-à-vis highly asymmetrical power relationships, since it is the role of trust within such relationships that appears to have received less attention within moral and political philosophy literature (a point to which we shall return later) (e.g., Baier 1986, 247–253).

We should note that a thorough examination of trust in connection with asymmetrical power relationships will inevitably make, implicit or explicit, reference to the notion of vulnerability. Indeed, any asymmetrical power relationship requires the involvement of at least one powerful and one vulnerable party. Insofar as this remains true, we can hypothesise that what trust does within asymmetrical power relationships is to mediate the relation between the powerful and the vulnerable party. The inclusion of vulnerability in an analysis of the relation between the notions of trust and power is further warranted by Annette Baier's approximation of trust as "accepted vulnerability to another's possible but not expected ill will (or lack of good will) toward one" (1986, 235). For Baier, in a trusting relation, the trustee leaves the trusted an opportunity to harm the trustee, and shows her confidence that the trusted will not exploit this opportunity. Here, 'an opportunity to harm' the trustee is a vulnerability of the trustee. Hence, how we think of trust partly depends of how we think of vulnerability.

In order to form an initial idea of how trust mediates the relations between the powerful and the vulnerable within asymmetric power relationships, we can consider again the episodic conception of power and vulnerability. On the episodic view of power, one person's power translates into another person's vulnerability. The more powerful is the former, the more vulnerable becomes the latter. This, however, presents us with a problem. Briefly stated, the more A is vulnerable to the power of B, the less likely A would be want to form a relationship with B, especially in the absence of certain checks on B's power or certain protections of A's vulnerability. Thus, the presence of power and vulnerability in society can become a barrier to the formation of enduring and cohesive social relations. This can also apply to human-technology relations: if a user feels that technology is likely to exploit the user's vulnerability, the user may wish not to use the technology in question. In such situations, trust can mediate between the powerful technology and the vulnerable user. Trust can be regarded as something that can delimit the extent to which power can be exercised, and by the same token, the extent to which one's vulnerability can be exploited. The mediating role of trust thus ensures the formation of cohesive social, institutional, and technological relations in society. The problem thus posed by relations of power and vulnerability is essentially the problem of social order as theorised by social contract theorists.

To understand how trust can mediate social relations between the powerful and the vulnerable, let us again consider the problem of social order in the state of nature (which we discussed briefly in Chapter 5). As Hobbes (1839) imagined, the state of nature is one of perpetual fear, struggle, and social disorder engendered by the ever-shifting and unstable relations of power and vulnerability. In the state of nature everyone is vulnerable to everyone else's power. This also applies to the powerful, since even the most powerful have their moments of vulnerability, for example, when they are asleep or taken ill (a similar point concerning the vulnerability of the powerful has been made in Section 8.2). Hobbes's own solution to the problem of social order in the state of nature consisted of the idea of a mutually binding social contract whereby warring individuals transfer their powers to a single sovereign—who can be a person or a group of persons—whose purpose would then be to direct individual actions towards the common benefit, if necessary, through coercion or force.

Despite its theoretical elegance and appeal, Hobbes's solution to the problem of social order has been subjected to much criticism. Consider one of the latest of such critiques. For example, Barry Barnes (1988, 22–24), drawing on the thought of Talcott Parsons (Parsons 1958), argues that the individualistic premises of Hobbes are very problematic. Specifically, Barnes has argued that Hobbes conceives of people in the state of nature as individualistic and rational egoists who may already have social institutions, such as common language or private property, which would presuppose the existence of some form of social order. If there are indeed social institutions, then the rational egoists depicted by Hobbes are already social actors. A similar critique of not only Hobbes, but the entire social contract tradition, has been advanced by John Searle (2010). The problem with the social contract theorists, according to Searle, is that they do not take proper account of the role of language in the constitution of society. The social contract theorists, as Searle notes,

> assume the existence of us as language-speaking creatures, and then they speculate how we might have got together in "a state of nature" to form a social contract. The point I will be making, over and over, is that once you have a shared language you already have a social contract; indeed, you already have society. If by "state of nature" is meant a state in which there are no human institutions, *then for language-speaking animals, there is no such thing as a state of nature.* (Searle 2010, 62; Searle's own emphasis)

Although Barnes deploys Parsons to critique Hobbes as shown above, Barnes finds Parsons's own sociological solution to the problem of social order ultimately unsatisfactory. On Parsons's sociological solution, social order emerges as a result of human socialisation through internalisation of social rules and norms. However, Barnes rejects this account of social order as resulting from rule-following. Barnes argues that if social life and order were indeed defined by norms and rules, "the rule-books would overstock any library and take more than a lifetime to read" (Barnes 1988, 27). According to Barnes, the most problematic feature of Parsons's account of social order is Parsons's unspoken acceptance of Hobbes's conception of people as egoistic and antisocial beings who need socialisation:

> Parsons' account of child socialization involves the social being, as it were, injected into the individual. Enter baby: and the rush is on to get him socialised before he is big enough and strong enough to embark upon a career of pillage, rapine and murder. Unfortunate egoistic tendencies have to be subordinated to social rules and standards. (Barnes 1988, 33)

By adopting a stance that stands closer to Searle than to Parsons, Barnes argues that human infants are not antisocial egoists who require to be disciplined and socialised. As with their characteristic predisposition to language acquisition, the pre-social infants exhibit innate tendency and capacity for socialisation (Barnes 1988, 33). In a way resonant with the analysis of habitus by Bourdieu and that of practical consciousness by Giddens (Haugaard

1997), Barnes contends that these predispositions reveal themselves in an innate desire for habituation and routinisation (1988, 34), and are, in the words of Haugaard, connected to "a predisposed trust that others will act in a predictable manner" (Haugaard 1997, 26). On this view, then, what holds society together is trust.

The forgoing discussion suggests two interrelated points. First, it suggests that relations of power and vulnerability in society are mediated through trust. Barnes, as noted above, views trust as an innate, or preprogrammed, social characteristic of human infants, which is akin to our predisposition to acquire language. This view of trust is to some extent analogous to the conception of trust developed by Annette Baier (1986), whose paradigm example of trust was an infant's trust in his or her parent (1986, 240–244). Baier used this example with the aim of shifting attention away from the contractarian views of trust as something that holds between free individuals of equal power, towards a "non-contract-based" view of trust that takes account of the unequal relations of power, dependence, and vulnerability in society (1986, 241). Baier thus was primarily interested in the mediating role of trust within relationships characterised by asymmetries of power and vulnerability; that is, the kind of relationships that have been overlooked by the contractarian accounts of morality. "A complete moral philosophy", Baier argued, "would tell us how and why we should act and feel toward others in relationships of shifting and varying power asymmetry and shifting and varying intimacy" (1986, 252).

Second, Barnes's critique of Hobbes and Parsons suggests that the nature of the relationship between the sovereign/state and subjects/citizens is better understood as not one of social contract or consent but as one of trust. (On this view, people's actions are guided by the trust that others will act or behave in a predictable manner.) A similar point can be found in John Dunn's reconstruction of the political thought of John Locke (Dunn 1984, 1988). Dunn maintains that inattention to the role trust plays in society constitutes a significant shortcoming in political philosophy (1988, 73–79). In this context, Dunn is interested in reinstating Locke's view of how subjects transfer their powers to their sovereign, "whom the Society hath set over it self, with this express or tacit Trust, That it shall be imployed for their good, and the preservation of their Property" (Locke 1988/1690, 381). In Dunn's reinterpretation of Locke's thought (Laslett 1988, 114–117), the relation between the state and citizens is not one of contract or consent, but one of trust. In this way, Dunn offers what Hardin described as "an insightful and compelling reconstruction of the otherwise incoherent move to justify government from a supposed grounding in contract or consent" (Hardin 1991, 204).

That the relationship between citizens and the state is one of trust and not contract also appears to apply to certain forms of human-technology relationship. Consider, for example, the process of signing up to a web-based service or application as a new user. As is known from experience, new users are invariably required to agree or consent to the so-called 'terms of service' or 'terms and conditions' which can be defined as "general and special arrangements, provisions, requirements, rules, specifications, and standards that form an integral part of an agreement or contract". While an act of agreeing to such 'terms' and 'conditions' can be interpreted as concluding a contract between us as new users and technology or service providers, rarely do we look into what these 'terms' and 'conditions' precisely say. We as ordinary users either do not comprehend the far-fetched, terminology-laden, legal texts of such terms and conditions, or we simply cannot be bothered to read them. In these instances, when we click on the 'agree' button, what we actually do is trust them.

7.5. CONCLUSION

Society is characterised by relations of power and vulnerability, which can be problematic in the sense that the more someone feels vulnerable in relation to others (whether these others are people, institutions, or technologies), the less likely one will be willing to form stable and enduring relations with these others. It is indeed plausible to expect that the vulnerable, other things being equal, tend to avoid the powerful to the extent possible. In this way, relations of power and vulnerability can give rise to the issue of social, institutional, and technological cohesion in society. There are several complementary ways of addressing the problem of cohesion between the powerful and the vulnerable. For example, the powerful can be made to be more *responsible* so that they do not abuse their power and be more considerate in their actions and treatment of the vulnerable. On the other hand, it is also possible to empower the vulnerable making them less vulnerable to the power of others. In this particular regard, *authenticity* can be a measure of the levels of empowerment of persons. Also, it is possible to establish *trust* between the powerful and the vulnerable as a means of mediating the relationship between the two parties. According to this somewhat simplified picture, the concepts of responsibility, vulnerability, authenticity, and trust can be regarded as a family of normative concepts involved in the pursuit of social, institutional, and technological cohesion.

Power, according to Steven Lukes, is a "primitive" concept, meaning that its sense "cannot be elucidated by reference to other notions whose meaning is less controversial than its own" (Lukes 2007, 83). This claim can also suggest that power as a primitive concept is presupposed by other less primitive concepts. As this chapter has shown concepts such as responsibility and vulnerability imply the presence or absence of power in persons or groups to whom they are applied. Ascription of responsibility often implies that the responsible person or group had some power to bring about a certain out-come, or has power to bring about some future outcome. Similarly, attribu-tion of vulnerability frequently implies that the vulnerable person has no power or that she is vulnerable in relation to some other powerful person. On this view, the way we conceive of power can affect our conception of con-cepts such as the ones discussed in this chapter. Although our discussion here is understandably sketchy (for obvious reasons of space), it is hoped that it can form a basis for further exploration of the conceptual connections be-tween power and other ethical and normative concepts such as the ones discussed in the chapter.

Chapter Eight

Power and Ethics

Practical Implications

Our exploration of the role of power in the ethics of technology continues in this chapter. Whereas the previous chapter has examined the conceptual role of power in the ethics of technology, by considering the notional connections between power and some key ethical concepts, the present chapter sets out to explore some major implications of power for the ethics of technology in practice, by considering how power can play a role in the processes of making and enacting ethical and socially responsible decisions and policies about technology. The discussion will centre on Responsible Research and Innovation as an emerging model of governance of science and technology in Europe. Crucially, the discussion highlights, and is premised on, the idea that power and powerful institutions can influence and shape social and moral values so often required for guiding research and innovation processes in society today. This suggests that the questions concerning power are inevitable and may require a serious consideration of the roles of democratic practices and ethics expertise in shaping the frameworks for Responsible Research and Innovation.

8.1. SETTING THE STAGE: RESPONSIBLE RESEARCH AND INNOVATION

The current landscape of science and technology governance and policy in Europe is increasingly characterised by the emergence and coming to prominence of *Responsible Research and Innovation* (RRI). The term RRI refers to policy initiatives of national, regional, and international research funding agencies (e.g., the European Commission) to conform the processes and outcomes of research and innovation to the societal needs and values (von Schomberg 2013). RRI also refers to the field of academic discourse which has emerged around such policy efforts (e.g., Stahl 2012; Stilgoe et al. 2013; Oftedal 2014; Søraker & Brey 2014; Zwart et al. 2014; Koops et al. 2015; Simon 2017; Gurzawska et al. 2017).

RRI has historical roots in previous discussions of research ethics, technology assessment, corporate social responsibility, and in discussions of the ethical, legal and social aspects of research and innovation (i.e., ELSA: Economic, Legal, and Social Aspects) in areas such as genomics and nanotechnology (e.g., Zwart et al. 2014; Gurzawska et al. 2017). Also, Van Oudsheusden (2014) has pointed to earlier work directed at public engagement, which has been part of certain forms of technology assessment, anticipatory governance, and other approaches aimed at making science more democratic and increasing public participation and deliberation. In Europe, the emergence of the RRI discourse is closely tied to the adoption of research funding initiatives promoting RRI both at national and EU levels. The European Commission in particular has proposed RRI as a strategy to shape research and innovation practices within the EU Framework Programme H2020 (European Commission 2012).

According to von Schomberg, RRI "should be understood as a strategy of stakeholders to become mutually responsive to each other, anticipating research and innovation outcomes aimed at the 'grand challenges' of our time, for which they share responsibility" (2013, 51). Furthermore, von Schomberg defines RRI as "a transparent, interactive process by which societal actors and innovators become mutually responsive to each other with a view to the (ethical) acceptability, sustainability and societal desirability of the innovation process and its marketable products (in order to allow a proper embedding of scientific and technological advances in our society)" (von Schomberg 2012, 50). In recent official statements by the European Commission, RRI has been described in terms similar to the definition of RRI given by von Schomberg: "RRI is an inclusive approach to research and innovation (R&I), to ensure that societal actors work together during the whole research and innovation process. It aims to better align both the process and outcomes of R&I with the values, needs and expectations of European society".[1]

Historically, technological innovations, as von Schomberg (2013) observes, were normally regulated or controlled by some central agent, with the state often serving the role of this agent. Modern technologies, he further notes, are distributed through market mechanisms, whereby the state demands from industry and market representatives that they should address the "three market hurdles" of "efficacy, quality and safety", before they can lawfully market their products and services (2013, 53). In this historical context, RRI appears to have emerged as an alternative mode of science and technology governance and policy, inspired, to some extent, by the realisation that the traditional approaches to research and innovation cannot sufficiently address the incorporation of societal values and needs in research and innovation processes (e.g., Owen et al. 2013, 32). Hence, RRI is characterised by a decisive shift toward socially inclusive and democratic approaches to science and technology governance by increasing the diversity of actors in research and innovation in a continuous and iterative fashion.

According to the official European Union (EU) interpretations (European Commission 2012), RRI as a policy initiative consists of six main elements, or "pillars", such as (1) citizen engagement and participation of societal actors in research and innovation; (2) ethics and compatibility of research and innovation processes and outcomes with fundamental values; (3) promotion of science literacy and science education in society; (4) attainment of gender equality in research and innovation processes and their content; (5) promotion and establishment of open access to scientific knowledge; and (6) transparent and accountable multi-stakeholder governance of science and technology (European Commission 2012).

In addition to the six pillars of RRI emphasised in policy discourses, Søraker and Brey (2014) have identified six dimensions of RRI emphasised in academic literature, namely, (1) a proactive approach towards research and innovation that benefits societal needs; (2) a consistent and continuous involvement of society in research and innovation processes; (3) anticipation of impacts, risks, and benefits, and reflection on societal needs and values; (4) transparency of the processes of research and innovation, as well as of the agendas of all stakeholders; (5) instituting collective and distributed models of responsibility for research and innovation which includes all stakeholders involved in research and innovation; and (6) establishing multi-stakeholder governance models with a focus on stakeholder engagement in research and innovation (see also Stilgoe et al. 2013; Taebi et al. 2014). These dimensions tend to be compatible with the EU interpretation, and therefore can be subsumed under the dimensions of stakeholder engagement, ethics and governance (e.g., Gurzawska et al. 2017).

The following sections consider some common issues and challenges that power can pose to the processes of making and enacting decisions and policies about technological innovation. In doing so, the discussion focuses on what can be considered as three main pillars of RRI—stakeholder engagement, ethical reflection, and societal values. The goal is not to critique particular policy or academic interpretations of RRI but to catalogue some main power-related problems that can emerge in the course of implementing RRI practices and activities, and of which policy makers, practitioners, ethicists, and other stakeholders should be aware. It should be borne in mind that RRI has not emerged out of thin air. The questions concerning the incorporation of social and moral values in the design of technology have long been the concern of the politics and ethics of technology (Jonas 1974; Unger 1994; Winner 1986, 1995; Sclove 1992, 1995; Bijker 1993, 1995; Feenberg 1995; Zimmermann 1995), including other areas of research and practice, such as Value-Sensitive Design (Friedman 1997). Neither does RRI exist in some political economic vacuum that shields it off from the reality of power politics and entrenched economic interest. Insofar as RRI is implemented in the political economic space characterised by powerful institutions, questions concerning power are inevitable. The aim of the remainder of this chapter is to catalogue as many such issues as possible.

8.2. POWER AND STAKEHOLDER ENGAGEMENT

Technology design often takes place within fragmented institutional and professional silos, where little consideration can be given to societal needs and values. One of the commonly cited cases is about technologies that end up reflecting the cultural worldview or economic interests of the designers more than the needs and values of the users for whom they were developed in the first place (e.g., Michelfelder et al. 2017). This can especially become commonplace with regard to the more marginalised, oppressed, or otherwise vulnerable segments of the population who through powerlessness of their own may have little influence or say in the design and development processes (Clarkson et al. 2003). In this respect, stakeholder engagement and participation can be a force for the democratisation of the processes and outcomes of research and innovation. When inspired by the normative political ideals of democracy, equality, or egalitarianism, stakeholder engagement and participation aim to give fair and equal weight to the needs, values, and interests of individuals, stakeholders, and other societal actors. One could thus argue that underlying recent calls for greater stakeholder engagement and participation is the fundamental respect for people and their capacity for self-government.

In philosophical literature, democracy sometimes is justified or evaluated in two different ways (Christiano 2018). On the one hand, there are those who favour democracy as a means to some other end, by comparing the outcomes of democratic modes of decision-making with those of other forms of decision-making (e.g., Mill 1991, 74; Dewey 1937; Sen 1999, 152). On the other, there are those who justify democracy as end in itself, by claiming that democratic forms of decision-making are good or desirable independent of the outcomes of employing them (e.g., Gould 1988, 45–85; Singer 1973, 30–41). Just like democracy, stakeholder engagement can, in principle, be justified on instrumental or intrinsic grounds. However, in literature on democratic approaches to technology, one encounters primarily instrumental justifications for greater stakeholder engagement and participation (e.g., Sclove

Power and Ethics 141

1995; Laird 1993; Bijker 1995). For example, Sclove maintains that democratic approach to technological decision-making "expresses and develops
individual moral freedom" (Sclove 1995, 35), while regarding 'freedom' as a
higher order ideal. The democratic approach to technology design thus
understood hopefully avoids imposing a certain conception of the good on
stakeholders.

Stakeholder engagement and participation can come in a variety of forms,
ranging between the weakest and the strongest levels of citizen participation
and engagement. These different levels of stakeholder engagement can be
illustrated using the concept of "participation ladder" which was originally
developed by Sherry Arnstein (1969) in the context of housing planning in
the United States in the 1960s, in order to conceptualise different levels of
participation of citizens in the planning processes. The participation ladder
consists of eight steps, such as citizen control, delegated power, partnership,
placation, consultation, informing, therapy, and manipulation. These seven
steps, Arnstein then places into three main categories, namely, (1) citizen
power, (2) tokenism, and (3) nonparticipation. The three categories of citizen
participation thus range from the most active participation (i.e., citizens being the drivers of innovation as co-actors) through the less active (i.e., citizens contributing their views on innovation as consultants) to the least active
(i.e., citizens being objects of action and manipulation by others) forms of
engagement and participation. As can be seen, the idea behind the ladder of
participation is essentially about measuring the extent of power and influence
to determine the outcome of a participatory process. Despite presenting a
simplified picture of stakeholder engagement processes, the ladder of participation reminds us that not all forms of stakeholder participation can be empowering. The concept can be used in implementing and evaluating different
forms of stakeholder participation in research and innovation processes with
a critical eye on the relationships of power which constitute such processes.
In recent years, the concept has been further developed and applied to different contexts of technology design (e.g., Connor 1988; Fung 2006; Tritter &
McCallum 2006; Östlund 2015).

Additionally, stakeholder engagement and participation can be organised
in accordance with two different conceptions of (deliberative) democracy:
procedural and substantive (Gutmann & Thompson 2004). The *procedural*
view of democracy maintains that principles governing democratic practices
should only include procedural principles that apply to decision-making process, such as the right to vote, equal suffrage, majority rule, and so on, that is,
principles that prescribe the manner in which decisions are made, without
prescribing the substance or content of these decisions, policies, or laws. On
this view, a decision or policy is good, right, or just only by virtue of having
been produced in some procedurally correct fashion. In contrast, the *substantive* view democracy contends that procedural principles on their own are

insufficient, given that simply following the procedures can produce unjust outcomes, for example, discrimination against minorities. The substantive view, therefore, insists that, in addition to procedural principles, democratic principles should also include substantive principles, such as the equality of opportunity, or right to basic healthcare (Gutmann & Thompson 2004, 23–24).

This distinction between procedural and substantive views of democracy is relevant to the organisation of stakeholder engagement, at least insofar as the process of stakeholder participation can be either structured in accordance with (a) the procedural approach, whereby an equal voice is given to everyone, experts and non-experts alike, or based on (b) the substantive approach, whereby the process of decision-making includes certain prerogative (ethical) values, norms, or precepts that will prescribe, to some extent, the substance or content of the decisions and policies to be made. The choice of a particular approach largely depends on what we think of the capacity of stakeholders to make their own decisions about science and technology. Perhaps all stakeholders possess sufficient knowledge or expertise to make their own decisions, because either there are no objective truths about the issues under consideration, or everyone is an expert in their own way, in one or the other domain, concerning the issues being negotiated. However, we should note that in the age of complex technologies, making decisions about research and innovation increasingly requires both social-ethical and scientific-technological expertise and knowledge (e.g., Jonas 1974). Therefore, some form of substantive-democratic stakeholder participation can be necessary.

Stakeholder participation also runs the risk of becoming a mere process of negotiation of private economic or political interests, a process in which the debate is not about what rationality requires, but about whether the interests of participating social groups are satisfactorily represented. Moreover, public and stakeholder engagement processes can become hijacked by powerful corporate lobbyists as has been well documented in research on democratic approaches to innovation (Flyvbjerg 1998). If there is no real commitment of the relevant social groups to go beyond their private interests and to engage seriously in a moral discussion about what technological designs are right and just, then the democratisation of research and innovation process might make more people content without adhering to any ethical ideals whatsoever. Although businesses, that occupy the design and production end of the supply chain, tend to be responsive to private interests of various social groups, and thus they might engage in practices that resemble the democratic stakeholder engagement and participation, it is very likely that such practices will fall short of being truly democratic, given that in the current capitalist and free-enterprise societies, businesses are more likely to pursue profit than to equally take into account all the interests of the relevant stakeholder groups.

Last but not least, stakeholder engagement faces the question of defining the very notions of 'relevant stakeholder' and 'societal actor'. As Langdon Winner (1993, 369–370) has argued, the notion of 'relevant stakeholder' is often regarded as unproblematic, for example, by social constructivists. Social constructivists do not answer the question of how 'stakeholders' should be defined, who decides whether or not a particular individual or group can properly be seen as representing a particular stakeholder group, and, for example, whether larger stakeholder groups should be given greater input in the deliberation process than smaller stakeholder groups. Democratic representation may therefore already be lost in the very process of deciding what the relevant stakeholders are (e.g., Brey 1998). "Who says what are relevant social groups and social interests? What about groups that have no voice but that, nevertheless, will be affected by the results of technological change? What of groups that have been suppressed or deliberately excluded? How does one account for potentially important choices that never surface as matters for debate and choice?" (Winner 1993, 369). The questions posed by Winner require careful reflection, because, as poststructuralist students of power would argue, attempts to delineate the relevant social groups and stakeholders are subject to the risk of excluding certain social groups and stakeholders. Put differently, power is exercised by defining who the relevant stakeholders should be.

8.3. POWER AND ETHICAL EXPERTISE

In addition to stakeholder engagement and participation, RRI emphasises the importance of ethical reflection and deliberation within processes of technological development and innovation. However, as noted earlier, greater stakeholder engagement understood as negotiation of stakeholder interests may fail to bring ethical reflection into these processes. For this reason, participatory approaches to research and innovation quite often require the inclusion of ethicists, whose expertise can then help inform democratic deliberations concerning technology (especially in matters of complex and uncertain risks). Ethicists serve as experts, for example, when they sit on public or academic panels, or when they are interviewed in media. When RRI is concerned, ethics expertise can help enrich the sources of expertise and viewpoints and add to the diversity of stakeholders in the processes of research and innovation. Ethics expertise is moreover required as part of ethics review committees performing ethical assessment of research and innovation projects. Nonetheless, as ethics expertise becomes part of research and innovation activities characterised by existing relations and disparities of power, it can become, in a several ways, a means to consolidating power relations and asymmetries within research and innovations settings.

First, ethics experts can become a means to seeking legitimacy of prede-termined political decisions. For example, in 2001, the German government established the German National Ethics Council, with the aim of facilitating public discourses concerning issues of medicine and biotechnology (Stafford 2005). However, the particular way in which the Council was established became the object of criticism, particularly in the media (2005). Specifically, some journalists criticised the council for being a political instrument of the then German chancellor Gerhard Schröder for manufacturing public accep-tance of political decisions that were predetermined before the actual deliber-ations of the council (Mali et al. 2012, 171). This case suggests that there is a risk of ethics experts becoming a vehicle for the legitimisation of political decisions reached behind closed doors.

Second, without proper social arrangements in making decisions (e.g., Devon & van de Poel 2004), inclusion of ethics experts in research and innovation processes can lead to the imposition of values of these experts on other stakeholders. The notion of "values imposition" has so far been most prominent in the field of bioethics. According to the standards set by the American Society for Bioethics and Humanities, ethical experts and consul-tants "need to be sensitive to their personal moral values and should take care not to impose their own values on other parties" (American Society for Bioethics and Humanities 2011, 9). It is further noted that such imposition of values can become a form of "moral hegemony" (2011, 7). It is therefore important that ethics experts should develop recognition of, and deference to, the diversity of moral values among stakeholders within research and innova-tion settings.

Third, we should also consider the possibility of conflicts of interest. The work performed by ethics experts is normally unpaid. One, therefore, would assume that such ethics experts do not have conflicts of interest. However, defining conflicts of interest purely in economic or financial terms unduly narrows our understanding of other potentially serious sources of conflict, for example, conflict of political interest. Within research and innovation set-tings, ethics experts may be required to deal with a plurality of different moral viewpoints (which may not be known in advance, see, e.g., Benford et al. 2015). In such cases, ethics experts may find it hard to resist what Hans-son has called "partisan ethical advocacy" (Hansson 2017), by either ex-pressing views or favouring sides that eventually blur the line between sub-jective opinions and objective application of moral theories to practical mat-ters.

Finally, we should also consider the extent to which ethics experts share responsibility for decisions reached with other stakeholders. "Sharing pow-er . . . for decisions . . . also means sharing responsibility" for those deci-sions; and when stakeholders including ethics experts "co-own" the conse-quences of collective decisions, "the boundaries of ethical accountability are blurred and fluid" (Frauenberger et al. 2017, 12). Such considerations are particularly important when certain stakeholders get the opportunity to influence collective decisions without themselves facing the negative outcomes of these decisions, for example, when the outcomes do not directly affect these stakeholders. Such cases, as Albert Bandura notes, can give rise to diffusion and/or displacement of outcome responsibility as a 'mechanism of moral disengagement' (Bandura 1999; 2016).

The above considerations suggest that one of the main challenges facing ethics expertise today is the question of how to delineate the appropriate role of expert ethical advice in the context of increasing transition toward participatory and inclusive democratic models of science and technology governance. We can best begin addressing this question by asking whether it is possible to become an expert in moral issues. József Kovács (2010) notes that "traditional, pre-modern societies" maintained that "there is a universally valid ethical canon and those who do not subscribe to it are simply mistaken". On this view, the notion of ethics expertise is unproblematic: "Those who knew the ethical canon, usually the religious authorities, were qualified to tell everyone else what is morally right and wrong, and others had better rely on their expertise" (Hansson 2017, 243).

The emergence of modern ethics expertise is closely related to the emergence of applied ethics as a distinct field of practice-oriented philosophical discourse. In its earlier days, applied ethics came under significant criticism (e.g., Noble et al. 1982), some arguing that moral expertise is not the same as moral wisdom (Baylis 1989). While this may be so, the claim to moral expertise today is much more modest. Although there is no "universal ethical canon, or a universally valid method", there is an emerging consensus that ethicists can be experts in identifying and applying ethical concepts to issues of public interest. Their expertise is not so much in providing conclusions or supplying answers to problems but in helping to clarify the connections between, and the outcomes of, different moral theories and viewpoints with regard to controversial issues (for a summary of tasks performed by ethics experts, see Elliott 2006).

Being expert in dealing with moral issues, therefore, "does not consist in furnishing answers, but rather in the deployment of skill in moral reasoning" (Adams 2018, 207). The role of ethics experts should be understood as that of being "the midwives of ethics, scaffolding and facilitating processes to form and negotiate ethos, rather than 'doing the ethics' on behalf of others" (Frauenberger et al. 2017, 13). The task of ethics experts is the provision of a "moral cartography" (Crosthwaite 1995) that can help navigate the complex terrain of ethical theories. This notion of moral expertise, incorporated into appropriate institutional structures can go a long way in furthering the goals of stakeholder engagement in research and innovation processes. When deployed in a thoroughly thought-through manner, ethics expertise can be a powerful means to challenging entrenched interests and power in this and other domains.

Appropriate implementation of ethics expertise in research and innovation processes is also required by the fact that the work of ethics experts can become reduced to what Frauenberger and colleagues have called "packaging", whereby the task of ethical reflection and deliberation within a research or innovation project becomes confined to a certain sub-task, or work package, which is then left to the ethics expert(s) on the team to be dealt with, without any input from the other participants of the project (Frauenberger et al. 2017, 12–13). In such situations, ethics can become disregarded as a bureaucratic formality, or an unnecessary burden that needs to be overcome in a 'tick-box' approach.

However, ethics does not need to be the exclusive domain of ethicists. Ethics expertise can, to some extent, be cultivated in certain societal actors, industry leaders, entrepreneurs, and so forth. In the current business and design culture, there are certain individuals and groups, such as corporate managers, engineers, and state actors, who can provide significant input into the process of technology development and implementation. Since these actors are in relatively more direct control of technology development process, they can be expected to come up with better, ethically responsible technology designs, implementations, and regulations (Brey 1998). Instilling moral values and precepts in technology designers, entrepreneurs, and industry leaders can be taken up, for example, by (higher) education, professional, and government institutions. Technological universities can make ethics education an integral part of their study programmes, by offering courses in ethics (e.g., business ethics, engineering ethics, ethics of technology). Professional societies and associations can also play a role in the ethics education of their members, by drawing up enforceable codes of professional ethical conduct (Unger 1994). Finally, governments can provide economic and other incentives for businesses and industries that engage in research and innovation in a responsible and ethical fashion (Gurzawska et al. 2017).

8.4. POWER AND VALUES

One of the main goals of RRI as a strategic initiative, as noted earlier, is the alignment of research and innovation processes with societal needs and values. Also, the central goal of both stakeholder engagement and ethical expertise, as discussed above, is to identify and clarify relevant societal needs and moral values, which then should guide the processes of technology development and implementation. In this regard, the task of identifying and analysing values, as well as their embedding and inscription in technology is of paramount importance to the effective implementation of RRI initiatives and activities. Given this consideration, RRI then requires, and can benefit from, concrete and effective approaches and methods that can help clarify and assess societal values, as well as embed these values in technology. While RRI can benefit from a number of such approaches (e.g., Value-Sensitive Design), the task of identifying and inscribing relevant values in technologies can be fraught with issues of power. The following subsections further elaborate on this point.

8.4.1. Values in Technology Design

The question concerning the role of values in the design and development of technology gained prominence with the emergence of value-centric approaches to the design of technology. One such prominent approach is known as *Value-Sensitive Design* (VSD). It refers to a philosophically grounded, and empirically informed, approach to the design of technological artefacts and systems which takes account of human and social values in a serious and systematic manner. The emergence of this approach is often associated with the publication of an edited volume titled *Human Values and the Design of Computer Technology* and authored by Batya Friedman and her colleagues (Friedman 1997). The contributed studies in this work were all based on the premise that social and moral values, whether deliberately or not, are inexorably incorporated in the design of technical artefacts and systems, and that a critical attention is required into these values and the manner in which they become inscribed or embedded in the design of these technologies (e.g., Simon 2017).

VSD then is an approach to the design and development of technologies that takes account of "human values in a principled and comprehensive manner" (Friedman et al. 2008, 69). As such, VSD involves a threefold methodology, consisting of conceptual, empirical, and technical inquiries. Conceptual investigations aim to ascertain the values implicated in technology design and the way in which competing values should be traded off during such design processes. Empirical investigations consist of observation, measurement, and documentation of the human and social context in which the technical artefact will be situated. Technical investigations can, on the one hand, focus on how existing technical properties can support or suppress certain human values, and, on the other, involve a proactive design of technologies to support the values identified in the conceptual stage of investigations. These three levels of investigations can be iterative, meaning that designers are allowed to modify their products in a continuous manner (Friedman et al. 2008).

VSD shares certain similarities with a number of similar approaches, such as disclosive computer ethics (Brey 2000, 2010), reflective design (Sengers et al. 2005), values at play (Nissenbaum 2005; Flanagan et al. 2008), and values in design (Knobel & Bowker 2011), to name a few. As a matter of fact, the term VSD is frequently employed as an umbrella category comprising these other approaches to technological design and development (e.g., Simon 2017). These approaches suggest that engineers and designers can, to some extent, translate various stakeholder values and societal needs into material design specifications or software code and algorithms. Thus, one fundamental idea shared by all these approaches is that moral, social, and political values are inexorably imported, inscribed, or embedded into technologies, especially in the design and development stages of technology. Hence, the important question: Which or whose values should be embedded into a technology?

As its name suggests, a core notion in VSD is that of values. But what is a value? In academic literature, one can find different conceptualisations of values. For example, as Carl Mitcham (2005) notes, there can be economic, social scientific, and philosophical interpretations of the notion of value. Academic discussions and debates concerning the notion of value can furthermore revolve around (a) the intension and extension of the notion of value, (b) the differences between individual and collective values, and (c) the differences between values understood as feelings and those defined as norms (Mitcham 2005). With regard to VSD, Friedman and colleagues define values broadly as that which "a person or a group of people consider important in life" (Friedman et al. 2006, 349). The notion of 'value' in the context of VSD, as Judith Simon notes, refer "not to economic valuation, but to societal values, . . . values as socially shared judgments about what and who is how important" (Simon 2017, 221).

Therein, however, we encounter issues relating to power. Research and innovation activities and their outcomes are increasingly distributed across a context characterised by social, moral, and cultural fragmentation. In this diverse socio-cultural landscape, the task of determining which or whose values should guide research and innovation becomes increasingly challenging. "Values", as Judith Simon notes, "are always someone's values, the shared judgments of some group as opposed to another. . . . when values are inscribed into technologies . . . it is always a process of imposing one's view on others" (2017, 221). The issue of values imposition also relates to the earlier mentioned problem of defining which stakeholders are relevant in the course of technological development and innovation (see Section 8.2). When research and innovation activities occur on a global scale (e.g., Stahl 2012; Archibugi & Iammarino 2002), the problem acquires a global character.

8.4.2. The Moral Machine Experiment

To illustrate the preceding point, consider the case of autonomous cars. In 2013, the National Highway Traffic Safety Administration of the U.S. Department of Transportation issued a taxonomic scale ranking the degrees of autonomy of so-called autonomous vehicles (also known as 'autonomous' or 'self-driving' cars; Reese 2016). The proposed taxonomy consists of six levels and ranges from "no driving automation" (level zero) to "full driving automation" (level five). The taxonomy offers detailed definitions of the six levels of automation (SAE International 2016). Majority of cars on the roads today (at least at the time of the writing of this book) are on levels zero and one. For example, there are already cars that can perform tasks such as parallel parking. However, cars at level five of automation would drive without a human driver, and would perform tasks such as making swerves to avoid other cars, humans, or objects on their way. Performing more complex manoeuvres, such as swerving pedestrians, would require autonomous cars to make value judgements. Making such value judgements echo versions of the famous "trolley problem" thought experiment:

> Edward is the driver of a trolley, whose brakes have just failed. On the track ahead of him are five people; the banks are so steep that they will not be able to get off the track in time. The track has a spur leading off to the right, and Edward can turn the trolley onto it. Unfortunately, there is one person on the right-hand track. Edward can turn the trolley, killing the one; or he can refrain from turning the trolley, killing the five. (Thomson 1976, 206)

We can formulate a version of the trolley problem that involves an autonomous car: an autonomous car faces some sudden obstacle on its way, say a large and heavy lorry, should it swerve onto a pavement and kill pedestrians, or should it hit the heavy obstacle and kill the passenger(s) whom the autonomous car is transporting? How the autonomous car will react in such a situation will have to be determined in advance, by programming a set of algorithms expressing a particular value judgement, or moral algorithms, into the code of the autonomous car. This problem is increasingly presenting a challenge to both policy makers and car manufacturers, as the vision of fully autonomous cars nears reality. The main question is what we should tell the car about how to do distribute inevitable harm among people in such situations? As a matter of fact, there has recently been a host of studies and inquiries seeking to address the question of whether, how, and what ethics can be programmed into autonomous vehicles (e.g., Maurer et al. 2015; Nyholm & Smids 2016; Fournier 2016; Goodall 2016; Greene 2016; Gogoll & Müller 2017; Shariff et al. 2017).

Amidst this flurry of studies, one interesting approach began to emerge, associated, in particular, with the works of Edmond Awad, Iyad Rahwan, and Jean-François Bonnefon (Awad et al. 2018; Bonnefon et al. 2016). Their approach, which we can describe as social-experimental ethics, was based on the premise that before policy makers and car manufacturers decide what ethical programme to incorporate into autonomous vehicles, they should first "understand what decisions human drivers would make if they had the time to react" in situations such as the one described above. To understand this, Edmond Awad and his colleagues designed a simple online game called the Moral Machine (Awad et al. 2018; http://moralmachine.mit.edu). The game presents players with a variant of the trolley dilemma mentioned earlier: an autonomous car faces an obstacle on its path; it can either hit the obstacle or make a swerve and hit something or someone else. The game consists of several rounds, and each new round of the game presents a modified version of the trolley dilemma with different obstacles and people to be either hit and killed or avoided and spared.

The Moral Machine was launched in June 2016. In the two years following its launch, the game attracted worldwide attention and allowed its makers to collect 39.61 million decisions from 233 countries and territories. The collected data was then analysed with a view to understand what decisions human drivers would make if they had the little time to react in a situation such as the trolley scenario. The findings of the study were published in a paper in the journal *Nature* (Awad et al. 2018). By using geo-location technology, the scientists were able to identify the country of residence of the research participants. This also enabled them to identify three distinct "moral clusters" of countries with "homogenous vectors of moral preferences", which are broadly consistent with the influential cultural map of the world developed by Ronald Inglehart and Christian Welzel (2005): (1) the Western cluster including North America and Western Europe, (2) the Eastern cluster containing East Asian and Islamic countries, and (3) the Southern cluster consisting of the Latin American countries of Central and South America. The three clusters exhibit variance in the weight the players give to some moral preference. For instance, players in the Eastern cluster exhibited a stronger moral preference to spare older people and kill younger people than players in the Southern and Western clusters. Players in the Southern cluster showed pronounced moral preference to spare higher-status people and athletic people than homeless and fat people. The authors also observed that players in countries with strong rule of law (as indexed by the Rule of Law, Kaufmann et al. 2011) exhibited more pronounced preference to spare lawful pedestrians than those who cross the road illegally. Despite these differences between clusters, there has also been a universal pattern of utilitarian moral preference to spare groups rather than individuals. Also, the weak preference to spare pedestrians over passengers seems to have been shared by players from all three clusters. Finally, dogs were more likely to be spared than cats.

The Moral Machine experiment appears to be by far the largest study carried out on cultural variation and moral preferences for machine intelligence. Certainly, policy makers and car manufacturers should be cautious in drawing conclusions from the findings of the research, since, as the authors admit, most participants of the experiment have been male university graduates (Awad et al. 2018). Nevertheless, the study found unambiguous cultural variation in moral preferences. The study finds support in that the observed cross-societal variation corresponds to previously recognised cultural groupings (e.g., Inglehart & Welzel 2005). Moreover, as an inquiry in cultural determinism, this project is not new, and builds on an existing tradition of studies of cultural variations in ethical judgements (e.g., Henrich et al. 2001). Manufacturers and policy makers, according to Awad and colleagues (2018), should be at least aware of such variations in moral preferences exhibited by drivers and other stakeholders in the countries and societies for which autonomous cars would be manufactured. While public moral preferences may not become the sole determinant of the values to be embedded in the design of autonomous cars, the consumer willingness to purchase, and the public willingness to tolerate, autonomous cars will to some extent depend on the acceptability of the moral values and norms to be programmed into intelligent machines (Bonnefon et al. 2016).

As research and innovation activities and their outcomes are increasingly distributed throughout the globe (e.g., Archibugi & Iammarino 2002; Castells 2010a), the decisions and policies adopted concerning moral algorithms to be programmed into autonomous cars in country can have implications beyond its borders. For example, Germany is one of the biggest car manufacturing and exporting countries in the world. In June 2017, the Ethics Commission on Automated Driving at the German Federal Ministry of Transport and Digital Infrastructure (BMVI) published a set of ethical guidelines for the operation of autonomous cars, described as "the first guidelines in the world for automated driving" (BMVI 2017b). The guidelines consist of twenty propositions, with proposition 9 stating that "[i]n the event of unavoidable accident situations, any distinction based on personal features (age, gender, physical or mental constitution) is strictly prohibited. It is also prohibited to offset victims against one another" (BMVI 2017a, 11). The specified proposition expresses a clearly non-utilitarian, or deontological, moral principle.

It still remains to be seen how future autonomous cars to be manufactured in conformity to these (or some other similar) principles would fare in terms of export and marketability. However, within a culturally fragmented landscape of innovation and trade, car manufacturers might soon face some moral and policy dilemmas. On the one hand, if our hypothetical manufactures are to become responsive to the particular demands of customers in a foreign country, then they could be accused of double of standards. On the other hand, if the manufacturers upheld their own standards and policies concerning the machine ethics, and therefore become unresponsive to the cultural requirements of the importing country, then they would most likely be stepping onto the terrain of values imposition or moral hegemony. It is however also possible that issues in autonomous car design arising from diverse cultural differences and moral preferences would be overcome, for example, by means of signing a bilateral or international trade treaty setting the standards for the regulation of autonomous cars in the countries involved. Nonetheless, even in this scenario, the question remains as to which or whose values would or should prevail in determining the substance and content of such legal treaties. However, given some of the past high-profile precedents of mainly technology companies (often Western) acquiescing to the demands of foreign countries (often Eastern) in the areas such as online censorship, it is quite plausible that the future car manufacturers might follow suit with regard to moral algorithms.

8.5. CONCLUSION

To conclude the discussion, how we conceive of the different pillars or dimensions of current RRI practices is ultimately bound up with the issues of power, its construction, distribution, and exercise. Most authors within the RRI academic discourse would agree that effective stakeholder participation in the processes of research and innovation can be ensured through democratic institutions that give equal political weight to the needs, rights, and interests of all stakeholders. On this view, democratic institutions are required both as a foundation and setting for effectively carrying out research and innovation practices that aim for fair and just stakeholder participation—a point repeatedly emphasised in political economy literature (e.g., Halperin, Siegle & Weinstein 2005; Acemoglu & Robinson 2000, 2012). The task of delineating the contours of stakeholder engagement especially in relation to the role of ethical reflection—which are two of the foremost pillars of RRI— in processes of research and innovation throws up issues of power that pose challenges to the current policy and academic discourses concerning RRI. The goal of identifying and incorporating social and user values in the processes of research and innovation raises further issues, in particular those of value imposition in a world of moral pluralism. Questions such as who should count as a stakeholder, what should be the extent of stakeholder involvement, whether stakeholder participation should be structured procedurally or substantively, whose values should guide research and innovation processes—are all bound up with the question concerning power.

Note

1. https://ec.europa.eu/programmes/horizon2020/en/h2020-section/science-and-society.

Chapter Nine

Power in the Absence of Ethics

Political Economy

We have reached the last chapter of our discussion of the relations between power, technology, and ethics. While the previous chapter has examined major implications of power for ethical decision-making about technology, the present chapter considers the effects of power on decision-making about technology in the absence of ethical and democratic practices and institutions in society. Responsible Research and Innovation, a socially responsible model of science and technology governance discussed in the preceding chapter, presupposes and requires the existence of suitably democratic structures and institutions for its implementation and operation. But what about countries or societies without such structures and institutions? To get an understanding of how research and innovation processes become governed in societies lacking democratic institutions, this chapter proposes to make a foray into recent political economy scholarship where there has been a growing interest in the role of economic and political institutions in determining the trajectories of technological development and innovation. The discussion aims to show that when power reigns free in the absence of ethics and social responsibility, different paths of technological development can become beset by inequalities of political power and the tyranny of entrenched economic interest.

9.1. POWER, INSTITUTIONS, AND TECHNOLOGY

The previous chapter considered the implications of power for the development and implementation of Responsible Research and Innovation (RRI) as a democratic, ethics-driven, and socially responsible framework for guiding research and innovation processes in Europe. We have seen that RRI, in its call for socially responsible practices and reflection on social values in processes of research and innovation, requires an adequately democratic institutional framework which can ensure that stakeholders and policy makers adopt and enact socially responsible decisions and policies on science and technology. However, there are countries which may lack such institutions and practices. Therefore, to understand (at least partially) how technology can be governed in the absence of socially responsible institutions we should look beyond the current European practices and consider those past and present cases in which technological development and innovation have been mainly left to political and economic power and institutions. Some recent research in political economy can help us in this regard, insofar as there has been a growing interest in the role of political and economic institutions in determining the varying paths of technological development in different countries.

In particular, since the publications of the influential works of Douglass North (North 1981, 1990, 1999), political economists have increasingly focused on the "right institutions" as being crucial factors influencing technological change and development (see also Nelson 2008). As part of this institutional trend in political economy research, Daron Acemoglu and James Robinson (2000, 2012) have recently put forward the "political-losers" model, according to which, in order to understand processes of technological change and innovation in different societies, it is vital to consider the nature of political and economic institutions and the distribution of political power in those societies. One good reason for considering political and economic power and institutions in relation to technology is that the development and implementation of technologies that best serve public interest takes place within the political and economic space which can significantly affect the making of decisions and policies concerning technological development and innovation. Put differently, action to be guided by practical reason occurs within a finite space of territorially bounded states, and can be subject to different legal and normative jurisdictions that can have different priorities that ensuring the development and adoption of technologies that best serve public interests (for more on this point see Section 9.4).

9.2. POLITICAL LOSERS AND TECHNOLOGY

To begin our discussion of the political-losers model as proposed by Acemoglu and Robinson (2000, 2012), consider the case of the diffusion of railways in nineteenth-century Europe and the United States. Although railways are generally considered as a key contributing factor to the Industrial Revolution and can serve public interest in terms of transport and communication, there were significant gaps and delays in their diffusion from an international comparative perspective. Thus, for instance, while in 1850 the United States already had 14,518 kilometers of railway tracks, Tsarist Russia, a country of comparable geographical size to the United States, had only 501 kilometers (Mitchell 1993). Similarly, while Britain boasted 9,797 kilometers of railway tracks in 1850, Austria had seven times less—merely 1,357 kilometers (Mitchell 1993). The question that needs addressing is this: Why do certain countries and societies, as in the above cases, fall short of adopting the best available technology which promises power to the state and prosperity to the people? It is this question that the political-losers model aims to address.

According to some economists who have dealt with this question, the adoption of novel technologies can become blocked by powerful interest groups who seek to safeguard their economic profits against potentially disruptive technological innovations. Thus, for example, a monopolist group can have an incentive to oppose and block the introduction of technological innovations by a rival group in order to protect its market monopoly. This economic model of resistance to technological innovations, according to Acemoglu and Robinson, has been quite prevalent in the economics literature and has been discussed by some prominent economists and historians (e.g., Kuznets 1968; Mokyr 1990; Grossman & Helpman 1994; Krusell & Rios-Rull 1996). Nonetheless, Acemoglu and Robinson argue, there are at least two problems with this "economic-losers" model. First, historically, there have been very few cases in which major technological innovations have been blocked by economic losers. As discussed by Mokyr (1990), even in the well-known case of the Luddites, skilled weavers whose trade and sources of income were threatened by the mechanisation of textile production, the economic losers were ultimately unable to stop the technological innovation of textile production. The second problem with the economic-losers model, according to Acemoglu and Robinson, is that it rests on the assumption that the economically losing groups possess the political power to block the technological innovations that threaten their economic interests. But if this is the case, Acemoglu and Robinson argue, "why not use this power to simply tax the gains generated by the introduction of the new technology?" (2000, 126).

Certainly, there can be limits to this political power. For example, the political power might be sufficient for blocking the technological innovation, but insufficient for taxing the gains of the technological innovation. Nevertheless, it does seem very plausible that an entity possessing sufficient political power to block the innovation would also be able "to lobby effectively for redistribution" (Acemoglu & Robinson 2000, 126) of the economic gains of the technological innovation through taxation. Thus, according to Acemoglu and Robinson, a more important reason for resisting technological innovations can be that the introduction of new technology can have both economic and political consequences, and thus affect not only the distribution of economic revenue, but also that of political power. In other words, as a better alternative to the existing "economic-losers" model the authors propose the "political-losers" model, according to which, technological advances will be blocked by those groups who fear that technological innovations will erode their political power. Indeed, if actors are economic losers without political power, then they will not be able to impede technological advances. If actors are economic losers with political power, then they have no incentive to hamper technological advances, since they can use their political power to redistribute the economic benefits resulting from the technological innovation to offset their economic losses. Therefore, Acemoglu and Robinson conclude, it is those actors who possess "political power and fear losing it who will have incentives to block" (Acemoglu and Robinson 2000, 126) technological advances. Their political model furthermore suggests that in order to understand technological backwardness, it is important to attend more to the nature of political and economic institutions and the distribution of political power in society (Acemoglu and Robinson 2000, 127; 2012).

By applying the political-losers model, Acemoglu and Robinson offer an interesting interpretation of why landed elites in Russia, but not in Germany, decided to oppose and block the railroads in the nineteenth century. The authors hypothesise that landed elites, who controlled political power just before the Industrial Revolution, resisted to the emergence of manufacturing in societies in which their political power was threatened, such as in Russia, but not in countries in which they could retain their political power, such as in Germany. Indeed, in Russia, a country ruled by an absolutist monarchy at the dawn of the Industrial Revolution, the landed elite blocked industrial innovations since they regarded such innovations as a threat to their political power. Industrial innovations were thus opposed because, "it was understood that industrial development might lead to social and political change" (Mosse 1992, 55). However, the situation was different in Germany, where the landed elite did not oppose and block industrialisation. As Acemoglu and Robinson (2000) note, the German landed elites formed the coalition of "Iron and Rye" with the emerging industrial elites in order to safeguard their economic interests, a coalition described as "a compromise between modern industry and the feudal aristocratic groups in the country" (Gerschenkron 1943, 49). In this way, the German landed elites gained political protection for their economic interests, shielding themselves from the undesirable consequences of the industrial innovation. All in all, according to Acemoglu and Robinson, it appears that the main difference in elite's attitudes between the two countries was the threat that industrial innovation posed to political power but not to economic interests.

To sum up, the political-losers model distinguishes between two kinds of resistance to technology: (1) cases in which technologies are simply opposed, and (2) cases in which technologies are effectively blocked. Given this distinction, the authors argue that while the economic-losers model can only explain why certain groups would oppose innovations (because they have economic interests in doing so), the political-losers model can further explain how certain groups can effectively block innovations (because they have sufficient political power to do so). Thus, the authors maintain (1) that economic losers *lacking* political power can merely resist technological innovations without effectively blocking them, and (2) that economic losers *possessing* political power would use this power to redistribute the benefits resulting from those innovations in their own favour, instead of resisting those innovations. In this manner, the political-losers model offers an interesting insight in how distribution of political power can affect processes of technological innovation. The model furthermore suggests that in order to understand technological backwardness of certain societies, it is necessary to begin by examining the nature of political institutions and the distribution of political power in those societies.

The research carried out by Acemoglu and Robinson is commendable in more than one respect. Firstly, the authors have fruitfully applied institutional thinking to matters of technology and innovation, by extending the more familiar terms and concepts of political power and institutions, thereby creating an interdisciplinary bridge to other areas of social and political inquiry which often use the same conceptual vocabulary. Secondly, the authors concern themselves with countries and societies beyond the developed global North and West, which tend to be frequently overlooked in much of current social and political studies of technology. Finally, their conceptually elegant model for explaining the varying technological paths taken by different societies is well supported with empirical data and illustrated by historical and contemporary examples. Nonetheless, in the following sections, we can further develop the political-losers model in two interrelated directions:

(1) The political-losers model appears to be applicable only to those cases in which the main variables are of economic and political nature. This focus on economic and political variables is quite understandable given that the model originates in the political economy scholarship. Nevertheless, as will become clear below, besides economic and political factors, there can be ideological factors that affect the processes of technological innovation. For this reason, it seems worthwhile to see whether the political-losers model can also be extended to situations in which technological innovations are resisted on ideological grounds.

(2) In its existing formulation, the political-losers model falls short of making explicit its conception of political power. Although Acemoglu and Robinson say much about the nature of political institutions (e.g., 2012, 42–44, 68–69, 79–87, 429–431), they do not provide a definition of political power (which is also missing from the index of key terms). The authors maintain that if political power is sufficient for blocking innovations, it should also be sufficient for

taxing the profits resulting from these innovations. Yet, they come short of making explicit precisely what it is about political power that makes it so effective in blocking technological innovation. Therefore, the model could be supplied with a suitable definition of political power.

The remainder of this chapter aims to further develop the political-losers model with respect to the two points expressed above. Thus, Section 9.3 extends the political-losers model to cases in which technology is resisted on ideological grounds by discussing the recent case of resistance to the liberalisation of the internet by the conservative clerical elites in Iran. Section 9.4 will take the discussion further by supplying a conception of political power befitting the political-losers model, which can help explain why political power can be a significant factor and an effective force in matters of technological innovation.

9.3. IDEOLOGY AND TECHNOLOGY

We can begin by noting that ideology, as a powerful source of values, norms, and beliefs, can affect processes of technological innovation by promoting or resisting certain technological designs and solutions. The term 'ideology' seems to have first been used in the late eighteenth century by Destutt de Tracy, who wanted to institute a "science of ideas" by applying methods of natural science to social phenomena (Kennedy 1979, 354). In the later period, however, the term acquired the familiar negative connotation of 'falsehood' and 'distortion', which still often persists to the present day (Steger 2007, 368).

Yet, falsehood is not all that there is to ideology. For example, French philosopher Paul Ricoeur (1986) identified three functions of ideology. The first function of ideology, for Ricoeur, is the distortion of images of social reality. Drawing on the works of Marx, Ricoeur postulated that the ideological distortion conceals the disparity between how things can be imagined in theory and how things play out in reality. Indeed, as Steger (2007) notes, ideologies present a picture of the world that both represents and distorts reality. Yet, for Ricoeur, unlike Marx, distortion is not all that there is to ideology (see also Tucker 1978). The second function of ideology, for Ricoeur, is the legitimation of authority. Drawing on the works of Weber (1946) and Mannheim (1936), Ricoeur characterised ideological legitimation as bridging the gap between people's belief in the legitimacy of the ruler and the ruler's claim to authority over people. Finally, the third function of ideology, for Ricoeur, is the integration of society. Drawing on the works of Geertz (1973), Ricoeur described the integrative function of ideology as the provision of values, norms, and beliefs that serve the processes of construction and maintenance of individual and social identity. This integrative function of ideology identified by Ricoeur, as Steger (2007) notes, echoes the idea of "hegemonic" ideology posited by Antonio Gramsci (1971), according to which, the ruling classes lure the working classes into accepting a shared identity that runs against their interests, thus making it possible for the ruling classes to maintain favourable social power relations without resorting to force and violence.

This threefold functional analysis of ideology by Ricoeur coheres with the characterisation of ideology as an institutional and systemic source of power by Michael Mann (1986, 1993, 2012, 2013) as has been discussed earlier in this work (see Subsection 4.2.1.). For Mann, ideological power can be understood here as the ability to bring about relevant social and political outcomes through appeal to shared norms, values, and beliefs. As such, ideological power stems from the human need (1) to give meaning to life; (2) to share values, norms, and beliefs; and (3) to take part in ritual and aesthetic practices with other humans (Mann 2013, 1–2). In this way, ideological power mainly involves the construction and communication of shared meanings, beliefs, norms, and values. Ideological power, according to Mann, is effective insofar as it can help people to attain some certainty or stability in their knowledge of external reality. In doing so, ideological beliefs fill in the gaps of uncertainty with "beliefs that are not themselves scientifically testable" but which embody or represent human fears and hopes (Mann 2013, 1). For example, it is either very difficult or wholly impossible to prove the existence of God in a scientifically testable fashion. Put differently, ideological beliefs are not rational or scientific beliefs, for as Maurice Bloch argued, "You cannot argue with a song" (Bloch 1974, 71; also cited in Mann 1986, 23). The construction and communication of ideological beliefs and meanings come as a response to developments in the spheres of economy and politics. Ideologies tend to be suddenly very important when humans have to wrestle with unexpected crisis in these economic and political realms. It is during such moments of crisis that we become "most susceptible to the power of ideologists who offer us plausible but scientifically or empirically un-testable theories of the world" (2013, 1).

We can observe a two-way relationship between ideology and technology. On the one hand, ideology can affect technological design and innovation processes (e.g., Noble 1984; Pfaffenberger 1990; Mackay & Gillespie 1992; Kozinets 2008). On the other hand, technology can embody different ideological values and norms (e.g., Tilley 1991; Myers 2001; Saunders 2000; Rowlands 2005). It would thus follow that ideologies can provide reasons for promoting or resisting technologies, depending on whether or not values embodied by those technologies conform to those expressed by the ideologies in question. Moreover, ideologies can provide resources or capability for resisting or promoting certain technologies in the form of ideological power. What is more interesting in this context is that the distribution of ideological power itself can be at stake as a result of technological innovation. Just as technological innovations can have an effect on the distribution of political power, as the political-losers model maintains, the distribution of ideological power can also be affected as a result of technological innovations. As discussed in more detail in a case study below, technological innovations can be opposed by interest groups, depending on whether these groups fear losing their ideological power and positions as a consequence of these technological innovations.

Let us now turn to an actual case in which a particular technological option appears to have been resisted on ideological grounds—the case of the recent attempts of the clerical conservative elite to block the liberalisation of the internet in Iran. In general, the net, with its relatively convenient, inexpensive, and anonymous avenues for sharing ideologically and politically sensitive information has long been a "digital dilemma" (Howard et al. 2011) for the authoritarian Iranian regime. Often leading the way in the Middle East in suppression of the internet freedom, Iran has frequently figured on the lists of "internet enemies" (e.g., Reporters Without Borders 2009). The situation with the freedom of the net particularly worsened in 2009, when, after the widespread protests against the conservative leaders, the authoritarian regime cracked down on media freedom, by banning online services such Twitter, Facebook, and YouTube, and in effect making many ordinary Iranians resort to using anti-censorship software (such as proxy servers and virtual private networks) in order to bypass internet censorship controls (BBC 2014; Morozov 2011).

Nonetheless, about four years after the events of 2009, there emerged some signs that the strict censorship regulations might eventually be eased to some extent. A first such sign came two weeks after Hassan Rouhani was elected as Iran's new president. Being a more moderate politician than his Holocaust-denying predecessor, Rouhani expressed his belief in the futility of net censorship as well as his intention to minimise censorship restrictions on the internet (Dehghan 2013a, 2013c). During a speech at the University of Tehran in 2014, Rouhani has also insisted that the internet is essential for connecting with the world of science: "We cannot close the gates of the world to our younger generation. . . . If we do not move towards the new generation of mobile today and resist it, we will have to do it tomorrow. If not, the day after tomorrow" (BBC 2014). A second notable sign was the fact that a month after the new president elect was sworn in, the administration of the newly elected president of Iran has embraced Twitter and Facebook, which has appeared as an open challenge to net censorship (Dehghan 2013b). Following Rouhani's suite, a number of cabinet members of president Rouhani—including the foreign minister, Mohammad Javad Zarif; the oil minister, Bijan Zanganeh; and even a spokeswoman for the Foreign Ministry of Iran, Marzieh Afkhami—have all become active online despite the fact that access to most of the major social networking services is blocked in the country (Dehghan 2013b). Finally, there emerged some tangible evidence supporting the intentions of Hassan Rouhani to ease internet restrictions. Soon after his election as the new president of Iran, Iranian internet users said that access to VPN accounts was restored (Dehghan 2013a). Another of the palpable changes since Rouhani came to power was an increase in the connection speed of the internet as reported by the Iranian net users (Dehghan 2013b; Hsu 2014). Those Iranians who were not able to access Skype because of the low speed of the internet became able to use Skype shortly after the election of moderates into power (Dehghan 2013b).

In this struggle to liberalise the use of the internet in Iran, it is possible to distinguish at least three main interest groups: (1) the conservative group: the clerical elites who are opposed to the liberalisation of the use of the net; (2) the moderate group: the incumbent political elites, who have recently expressed intent to liberalise the internet; and (3) the business group: the business elites, who are seeking returns on their investments in information and communication technologies (but who lack significant political power to secure their economic and financial interests).

The conservative group's resistance to the liberalisation of the net seems to have stemmed from their fears that it will negatively affect their ideological power. Clearly, for a group that derives its political power from conservative Islamic ideology, the internet that offers the possibility of freely and easily sharing and communicating ideologically and politically sensitive information is a significant threat. Conservative clerics in Iran particularly opposed the introduction of 3G mobile broadband services, arguing that they made it possible for immoral images to be shared more freely and easily. As Ayatollah Makarem Shirazi, a prominent conservative Iranian cleric, said, mobile internet was "immoral and unlawful" (BBC 2014). Indeed, hardliner clerics, who supported Ahmadinejad for presidency in the elections of 2005 and 2009, heavily relied on the votes of the conservative segment of the population (Ansari 2008). Given this reliance on conservative voters, erosion of conservative values presents a real threat to their political power which they derive from their ideological positions. Indeed, as noted earlier, while the conservatives were in power (i.e., between 2005 and 2013) the existing speed of the internet has been significantly lowered, and up to five million websites came under censorship (BBC 2014; Morozov 2011). Remarkably, for the conservative group, the liberalisation of the net would have ideological consequences which they cannot easily offset through taxation. Indeed, if the conservative group had economic interests in mind, instead of lowering the speed of the internet connection and blocking millions of websites, they would have simply taxed the gains resulting from the liberalisation of the net. Yet, the group chose censorship instead of taxation, primarily because the ideological losses resulting from the liberalisation of the internet could not be easily offset in monetary terms.

The conservative elite had more to lose, in comparison with the moderate elites, from the liberalisation of the said communication technology. The moderate group, unlike the conservative group, relies less on the conservative segment of population but more on the moderate liberal youth: indeed, in the aftermath of the 2009 presidential election, in which the moderates lost to the conservative candidate, it was primarily the liberal youth that came out in the streets to protest against the result of the election. Thus, in terms of ideological power, the moderate group has less to lose from the liberalisation of the internet, which partly explained their endorsement of the liberalisation of the media and the internet in Iran. Moreover, it can be argued that for the moderate group, there are as well some underlying economic interests in liberalising the net. Once, the moderates came to power in 2013, they granted 3G licences to at least three mobile phone firms (BBC 2014). Thus, moderates chose to seek rent from economic gains resulting from the granting of the 3G mobile broadband licences, instead of resisting the internet technologies and heavily censoring the traffic. Indeed, the speed of the internet traffic in Iran has deliberately been reduced in the aftermath of the election protests of 2009 in Iran (Hsu 2014). Hence, the lower speeds experienced by Iranians during the second presidential term of Mahmoud Ahmadinejad was not a result of lack of technological infrastructure that allows for higher connection speed, but a result of the crackdown on internet use. With the departure of Ahmadinejad from power, those who already had vested economic interests in internet-providing services prior to the protests of 2009, came out of the shadows and effectively lobbied Rouhani and his cabinet to ease internet restrictions. The moderates thus appear to have acted on the basis of the idea that it is better to remove the restrictions off the existing technological infrastructure and tax it, rather than resist it while the population seeks other means to circumvent the restrictions.

The existence of the business elite as an interest group can be inferred from the fact that most internet-providing companies are privately owned in Iran (Reporters Without Borders 2009, 13). This business group had been 'economic losers' ever since the hardliners came to power and through the crackdown on net freedom in the aftermath of the 2009 election. While the privately owned internet communication infrastructure was able to provide faster broadband internet, internet speed had been deliberately lowered during the rule of hardliners. Clearly those who have invested in the infrastructure would lose economically as a result of the lowering of the speed of the internet and thus seek returns on their investments. However, the business elite lacked failed to secure their economic interests as a result of lack of political power and inability to get political support from the government that was recently affected by the mass protest events of 2009 in which social media services played a significant role. Thus, lacking in political power and unable to secure their economic interests, the business elite could only bide their time until more favourable political conditions would set in. The moderate group who replaced the conservative group in 2013 provided exactly the kind of favourable conditions that the business elites had been anticipating. Unlike their conservative predecessors, the moderates, represented by Rouhani, were not intimidated by the net in terms of losses of ideological power. In this way, the moderate group chose to liberalise the internet and establish a proper presence in the digital realm, while also enjoying the economic benefits that would follow from the liberalisation of the net.

The forgoing case study was offered to inquiry whether the political-losers model can be extended and applied to scenarios in which technological innovations are resisted on ideological grounds. In its original formulation by Acemoglu and Robinson (2000), the political-losers model distinguished between two kinds of resistance to technology: between cases in which technologies are simply opposed and those in which technologies are effectively blocked. On the basis of this distinction, the authors have argued that while the economic-losers model can only explain why certain groups would oppose innovations (because they have economic interests in doing so), the political-losers model can further explain how certain groups can effectively block innovations (because they have sufficient political power to do so). The authors thus maintain (1) that economic losers lacking political power can merely express their opposition to technological innovations without being able to actually block them, and (2) that only those possessing political power can actually block technological innovations. Now the question that needs addressing is whether similar conclusions can be drawn about ideological losers.

Regarding the first thesis, it indeed seems to be the case that ideological losers lacking political power can only resist technological innovations without being able to actually block them. As we have seen, the conservative clerical elites (who resisted the internet on ideological grounds) became further removed from political power when their representative Ahmadinejad left office and Rouhani, the leader of the moderates, was sworn in as the president. During this more pro-business presidency of Rouhani, the conservative clerical elite could only voice their opposition to the liberalisation of the internet without being able to actually block such liberalisation. Thus, in this period, the conservative elites lacked political power to secure their ideological interests just like the business elites who had lacked political influence to secure their economic interests during the rule of conservative Ahmadinejad. This suggests that ideological power does not provide sufficient resources to block technological innovations unless such power is successfully converted into political power.

Correspondingly, the second thesis—that only those possessing political power can actually block technological innovations—does seem to be applicable to ideological losers as well. Thus, for instance, during the rule of conservative Ahmadinejad, the conservative clerical elite enjoyed more political power and influence. They were thus in a position to actually put significant restrictions on the internet and successfully block attempts to remove those restrictions while their representative was in power. Similarly, the business elites, who, during the rule of the conservatives, had failed to secure their economic interests, eventually came to enjoy a more favourable political environment and gained political support for securing their economic interests during the rule of the moderates. Again, we observe the important role political power plays in shaping the trajectory of technological development and innovation. The gains and losses of the interest groups in question can be given a simplified description as in Table 9.1 below.

The foregoing discussion thus shows that the core thesis of the political-losers model—that without sufficient political power interest groups can merely oppose innovations without being able to actually block them—can be extended and applied to scenarios where technological innovations are resisted on ideological grounds. Thus, on the extended version of the political-losers model, (1) there can be instances of technological innovation, in which technological innovations are opposed by interest groups who fear losing their ideological power and positions; and (2) while ideology can provide reasons for resisting technological innovations, it cannot provide resources for effectively blocking those innovations, unless ideological power is effectively tied to political power.

Furthermore, the extended version of the model can highlight an important difference between ideological and economic losers. As Acemoglu and Robinson (2000) note, economic losers possessing political power will not resist technological innovations but use their political power to redistribute

Table 9.1.   Ideological Winners and Losers.

Conservative Ahmadinejad ir power (2005—2013)	• Ahmadinejad's government place heavy restrictions on the internet; • Conservative elites are *ideological winners* as result of these restrictions; • Business elites are *economic losers* as a result of these restrictions.
Moderate Rouhani in power (2013—ongoing)	• Moderates lift media and internet restrictions to a significant degree; • Conservatives are *ideological losers* as a result of the liberalisation of the internet; • Business elite are *economic winners* as a result of the liberalisation of the internet.

the benefits of such innovations in their own favour. This thesis however might not necessarily be applicable to ideological losers, since, unlike economic losses, ideological losses cannot be compensated for through taxation in an easy and straightforward way. This is mainly the issue of "incommensurability" (Griffin 1997) or "incomparability" (Raz 1986, 1997) of values: certain ideological losses, which cannot be compared with, or evaluated in terms of, economic resources, cannot be easily compensated for in economic or monetary terms. Certainly, here one could adopt a pragmatist approach as a solution to the issue of incommensurability (e.g., Anderson 1997) by proposing indirect ways of offsetting ideological losses through taxation. For example, the ideologically conservative group might use monetary resources gained through taxation to offset their ideological losses in one area of activity by strengthening their ideological grip in other areas of activity, say, by building religious schools and places of worship. However, compensating for ideological losses in such an indirect way is not as easy and straightforward as compensating economic losses through taxation. In principle, it was possible for the conservative clerical elites to gain economically by promoting liberal policies with regard to the internet and media. However, championing such liberal values would also have significantly affected their ideological image among the conservative segments of the population. Thus, the conservative elites faced a version of the "digital dilemma" (Howard et al. 2011): whether to make a trade-off in favour of their ideological identity or potential economic gains. As we have seen above, the conservatives chose the former rather than the latter option.

The inability to offset ideological losses directly through taxation makes ideological interest groups less flexible and therefore more regressive when faced with a potentially disruptive technological innovation. In contrast, the possibility of offsetting economic losses through taxation makes economic interest groups more flexible in dealing with the effects of economically disruptive technologies. Understanding this difference in behaviour between ideological and economic interest groups can be useful, for example, in making predictions about the attitudes of politically powerful interest groups to technological change within a given society, in particular those powerful groups whose political power is based on either ideological or economic resources. According to some theorists (e.g., Giddens 1985a, 7–8; Mann 1986; Mead 2005; Nye 2011), sources of political power can be of ideological or economic nature. In a similar vein, political regimes can have ideological or economic bases of legitimacy (e.g., White 1986; Alagappa 1995). Now extending the earlier noted difference in behaviour between ideological and economic interest groups, it can be suggested that politically powerful groups whose power and legitimacy are based on conservative ideology can be more regressive in their attitude toward technological change since their ideological losses cannot be directly compensated for through taxation. This goes on to show that in considering resistance to technological change and innovation, it is vital to attend not only to the distribution of political power but also to the sources of political power.

9.4. THE STATE AND TECHNOLOGY

Overall, the political-losers model emphasises that to understand why different societies follow different technological trajectories, we should consider the nature of economic and political institutions and the distribution of political power in those societies. The model can therefore be deployed to explain the role of economic and political institutions in processes of making and enacting policies concerning technology, especially in the absence of democratic and ethical structures and institutions for guiding technological innovations. While the previous section has extended the political-losers model to cases where technology is opposed on ideological grounds, this section takes the discussion further by supplying the model with a suitable conception of political power, one that can explain precisely what it is about political power that often makes it so effective in blocking technological innovations. A step in this direction suggests that we identify and consider the sources, or bases, of political power.

Political power as an institution, at its very core, can be said to consist of the function of judicial regulation backed by coercion (and occasionally, if necessary, by force and violence). However, a function of judicial regulation also exists within other institutions of power, such as economic institutions (e.g., international investment arbitration), ideological institutions (e.g., religious courts), and military institutions (e.g., military tribunals that are independent from civilian courts). To distinguish political power from economic, ideological, and military forms of judicial regulation, it is necessary to restrict political power to judicial regulation backed by coercion and force which is "centrally administered and territorially bounded" (Mann 2012, 12). This means that political power is *state* political power, and it is this feature of political power—political power as exercised by one of the most political institutions of all—that makes it an effective force in matters of technological innovation.

There are certainly other conceptions of political power that do not necessarily tie it to the state. For example, political theorists often find 'political power', not just in states but also in a variety of organisations, in the form of governance administered by a miscellany of entities such as corporations, civil movements, non-governmental organisations, and so on (see, e.g., Mann 2012, 12). However, there are reasons to distinguish 'political power' understood as state political power from these other forms of organisational and bureaucratic power. As Mann convincingly argues, "States, not NGOs or corporations, have the centralized territorial form that makes their rules authoritative over persons residing in their territories. I can resign membership of an NGO or a corporation and so flaunt its rules. I must obey the rules of the state in whose territory I reside or suffer punishment" (2012, 12). On this view, political power as a concept is reserved for 'state political power', which mainly involves the centralised regulation of social life over a certain geographical territory (Mann 1986). As it stands today, there is indeed virtually no escape from the sovereign power of the state:

> [T]he worst fate that can befall a human being is to be stateless. . . . The old forms of social identification are no longer absolutely necessary. A man can lead a reasonably full life without a family, a fixed local residence, or a religious affiliation, but if he is stateless, he is nothing. He has no rights, no security, and little opportunity for a useful career. There is no salvation on earth outside the framework of an organized state. (Strayer 1973, 3)

Political power understood as state political power is thus both (1) territorialised in the sense of being used in the regulation of social life in geographically circumscribed area, whether at the local, regional, or national levels of governance; and (2) centralised in the sense of being the final authority or 'the last say' in settling relevant issues within a given territory. In this way, the political power of the state becomes strongly authoritative over the people residing in its territory. To see how the territorialised and centralised nature of state political power plays out in the processes of technological innovation, consider for instance the influence sovereign political power had in the earlier stages of the development of the internet. The network architecture of the internet was initially developed without any regard to physical geography; that is, as a deterritorialised network: "[I]n Cyberspace, physical borders no longer function as signposts informing individuals of the obligations assumed by entering a new, legally significant, place" (Johnson & Post 1996, 1375).

However, this architectural feature of the internet would soon be challenged by the sovereign political powers and interests, one of the most emblematic instances of which being the litigation between LICRA[1] and Yahoo! Inc. in France. In 2001, LICRA complained to Le Tribunal de Grande Instance of Paris (The High Court of Paris) that Yahoo! was allowing the sale of Nazi memorabilia on its auction website contrary to Article R645-1 of the French Criminal Code (Okoniewski 2002). In response, Yahoo! argued that there were no technological means of preventing French users from accessing these auctions (Okoniewski 2002). However, the impossibility argument put forward by Yahoo! would soon be quashed, when Cyril Houri, the founder of a start-up company called Infosplit, contacted and informed Stephane Lilti, the plaintiff's lawyer, about the existence of a newly developed technology that could filter out internet users on the basis of their geographical location (Goldsmith & Wu 2006). During a subsequent court hearing, in response to Yahoo!'s impossibility argument, the plaintiff's lawyer, Lilti, argued by noting the existence of Houri's geo-location technology and claiming that the firm, if it wanted, could filter out web users on the basis of their physical location (Goldsmith & Wu 2006). Thus, the French court ordered Yahoo! to eliminate French internet users' access to Nazi related materials on its websites, the order with which Yahoo! eventually had to comply (see Okoniewski 2002 for a detailed discussion of the case).

While the story of the litigation between LICRA and Yahoo! captures the ever-evolving transformation of the internet from a deterritorialised network technology to a technology that increasingly embodies physical and political geography, it also serves as an example of how powerful state political power can change and shape the course of development of particular technologies. Importantly, it shows how economic interests in technology can be quashed by sovereign political power. While Yahoo! might have thought that it would be impossible for the French court to extend its power into the U.S. jurisdiction, the company was in no position to escape the French law, since it had significant assets and a sizable subsidiary on the French territory at risk of seizure (Goldsmith & Wu 2006, 8; Reidenberg 2002, 269). Indeed, the French court warned the firm that it had to comply with its decision or face heavy fines (Okoniewski 2002). The decision of the French court marked the beginning of the trend of bringing the internet under the sovereign state political power in many countries of the world, the trend which Nocetti (2015) has described as "sovereignisation" in his recent article studying the Russian "dictatorship-of-the-law" approach to internet policy aimed at the gradual isolation of the Russian internet from the global infrastructure.

9.5. CONCLUSION

This chapter started by asking what implications power can have on processes of research and innovation in the absence of ethical and democratic practices and institutions in society. To address this question, we made a quick foray into recent political economy research to identify those economic, ideological, and political institutions that can become a powerful force in determining paths of technological development and innovation. Power, especially in its political form is one of the most striven for things. It is often more important than wealth, especially when the latter cannot buy the former and when the former—in the form of state political power—can help redistribute the latter. Given this frequent desirability of power over wealth, as the foregoing discussion has shown, in considering the different technological trajectories certain societies have chosen, it is important to begin by examining the distribution and the sources of political power in those societies. In coming to this conclusion, the chapter has built on the political-losers model of Acemoglu and Robinson, according to which powerful interest groups can not only oppose but effectively block those technological innovations that pose a threat to their political power. Following a discussion of the political-losers model, the chapter has further developed the political-losers thesis in two directions. First, the model has been extended to cases of ideological resistance to technology by discussing the recent case of resistance to the liberalisation of the internet by the conservative clerical elites in Iran. Second, the model has been supplied with a suitable conception of political power as 'state political power' which can help explain why political power can be a significant factor and an effective force in matters of technological innovation.

Note

1. LICRA stands for La Ligue Internationale Contre le Racisme et L'Antisémitisme (The International League against Racism and Anti-Semitism).

Conclusion

To sum up, the notion of technological power is fast becoming an object of both academic and policy discourses. Within such discourses one can observe several different senses of power being employed, especially when viewed from a multi-disciplinary angle. This demonstrates the need for the creation of a cross-disciplinary and integrative framework for identifying and clarifying different conceptions of the relations between technology and power. In this respect, the book aims to offer an empirically informed philosophical framework for understanding the technological construction of power, which allows for a differentiated vocabulary for describing various senses of technological power, while bridging together social and political theory, critical studies of technology, philosophy, and ethics of technology. In addition, the framework presented in this book aims to contribute to better critical and ethical evaluation of technologies and their powers. Any adequate ethics or critique of technology must be based on a better, clearer, and more nuanced and differentiated understanding of the many ways in which technology can be described as 'powerful'.

Perhaps the discussion in this book has only scratched the surface of the nexus between technology and power when compared to the greats of critical thought. However, the book would have done more than it has aimed for if it makes at least some contribution to the pluralist approach to power and technology. The overarching message has been that we need to recognise power in as many of its guises as possible given the vast field of multiplicity of forces and factors that characterise the reality of power. It is time for sectional and disciplinary differences to be set aside and for creating a united intellectual front at least for sake of the powerless, the vulnerable, the poor, and the marginalised.

References

Acemoglu, Daron, and James A. Robinson. 2000. "Political Losers as a Barrier to Economic Development." *The American Economic Review* 90, 2: 126–130.

Acemoglu, Daron, and James A. Robinson. 2012. *Why Nations Fail: The Origins of Power, Prosperity and Poverty*. London: Profile Books.

Adams, David M. 2018. "Are Hospital Ethicists Experts? Taking Ethical Expertise Seriously." In *Moral Expertise: New Essays from Theoretical and Clinical Bioethics*, edited by Jamie Carlin Watson and Laura K. Guidry-Grimes, 207–225. Cham: Springer International.

Aghion, Philippe, and Peter Howitt. 2009. *The Economics of Growth*. Cambridge, MA: MIT Press.

Akrich, Madeleine. 1992. "The De-Scription of Technical Objects." In *Shaping Technology/ Building Society: Studies in Sociotechnical Change*, edited by Wiebe E. Bijker and John Law, 205–224. Cambridge, MA: MIT Press.

Alagappa, Muthiah. 1995. "Anatomy of Legitimacy." In *Political Legitimacy in Southeast Asia: The Quest for Moral Authority*, edited by Muthiah Alagappa, 11–30. Stanford: Stanford University Press.

Allen, Amy. 1999. *The Power of Feminist Theory: Domination Resistance Solidarity*. Boulder, CO: Westview Press.

Allen, Amy. 2016. "Feminist Perspectives on Power." The Stanford Encyclopedia of Philosophy (Fall 2016), edited by Edward N. Zalta. https://plato.stanford.edu/archives/fall2016/entries/feminist-power.

Alpaydin, Ethem. 2010. *Introduction to Machine Learning*. Cambridge, MA: MIT Press.

Åm, Trond Grønli. 2011. "Trust in Nanotechnology? On Trust as Analytical Tool in Social Research on Emerging Technologies." *Nanoethics* 5, no. 1: 15–28.

American Society for Bioethics and Humanities. 2011. *Core Competencies for Health Care Ethics Consultation: The Report of the American Society for Bioethics and Humanities* (2nd edition). Glenview, IL: American Society for Bioethics and Humanities.

Amsden, Alice H. 1992. *Asia's Next Giant*. New York: Oxford University Press.

Amsden, Alice H. 2001. *The Rise of "The Rest": Challenges to the West from Late-Industrializing Economies*. Oxford: Oxford University Press.

Ananny, Mike. 2016. "Toward an Ethics of Algorithms: Convening, Observation, Probability, and Timeliness." *Science, Technology, & Human Values* 41, no. 1: 93–117.

Anderson, Elizabeth. 1997. "Practical Reason and Incommensurable Goods." In *Incommensurability, Incomparability and Practical Reason*, edited by Ruth Chang, 90–109. Cambridge: Harvard University Press.

Ansari, Ali. 2008. "Iran under Ahmadinejad: Populism and Its Malcontents." *International Affairs* 84, no. 4: 683–700.

Archibugi, Daniele, and Simona Iammarino. 2002. "The Globalization of Technological Innovation: Definition and Evidence." *Review of International Political Economy* 9, no. 1: 98–122.

Arendt, Hannah. 1958. *The Human Condition*. Chicago: The University of Chicago Press.

Arendt, Hannah. 1970. *On Violence*. London: Penguin.

Arendt, Hannah. 1973. *The Origins of Totalitarianism*. San Diego: Houghton Mifflin Harcourt.

Aristotle. 1998. *Politics*. Translated by C.D.C. Reeve. Indianapolis: Hackett.

Aristotle. 2006. *Metaphysics Book Theta*. Translated by Stephen Makin. Oxford: Clarendon Press.

Arnstein, Sherry R. 1969. "A Ladder of Citizen Participation." *Journal of the American Institute of Planners* 35, no. 4: 216–224.

Aron, Raymond. 1964. "Macht, Power, Puissance: Prose Démocratique ou Poésie Démoniaque?" *European Journal of Sociology/Archives Européennes de Sociologie* 5, no. 1: 26–51.

Ash, Timothy Garton. 1995. "Prague: Intellectuals and Politicians." *The New York Review of Books* 42, no. 1: 39.

Austin, John L. 1961. *Philosophical Papers*, edited by J. O. Urmson, and G. J. Warnock. Oxford: Clarendon Press.

Awad, Edmond, Sohan Dsouza, Richard Kim, Jonathan Schulz, Joseph Henrich, Azim Shariff, Jean-François Bonnefon, and Iyad Rahwan. 2018. "The Moral Machine Experiment." *Nature* 563, no. 7729: 59–64.

Ayers, Michael Richard. 1968. *The Refutation of Determinism: An Essay in Philosophical Logic*. London: Routledge.

Bachrach, Peter, and Morton Baratz. 1962. "The Two Faces of Power." *American Political Science Review* 56: 947–952.

Bachrach, Peter, and Morton Baratz. 1970. *Power and Poverty: Theory and Practice*. New York: Oxford University Press.

Baier, Anette. 1986. "Trust and Antitrust." *Ethics* 96, no. 2: 231–260.

Baker, Jim. 2018. "Artificial Intelligence—A Counterintelligence Perspective: Part I." *Lawfare*, August 15, 2018. https://www.lawfareblog.com/artificial-intelligence-counterintelligence-perspective-part-1.

Balkin, Jack M. 2016. "Information Fiduciaries and the First Amendment." *U.C. Davis Law Review* 49,1183–1234.

Ball, Terence. 1976. Reviews of *Power: A Radical View*, by Steven Lukes, and *The Descriptive Analysis of Power*, by Jack H. Nagel. *Political Theory* 4: 246-249.

Bandura, Albert. 1999. "Moral Disengagement in the Perpetration of Inhumanities." *Personality and Social Psychology Review* 3, no. 3: 193–209.

Bandura, Albert. 2016. *Moral Disengagement: How People Do Harm and Live with Themselves*. New York: Worth Publishers.

Barnes, Barry. 1988. *The Nature of Power*. Cambridge: Polity Press.

Barocas, Solon, and Andrew D. Selbst. 2016. "Big Data's Disparate Impact." *California Law Review* 104, 671–732.

Bartlett, Randall. 1989. *Economics and Power: An Inquiry into Human Relations and Markets*. Cambridge: Cambridge University Press.

Bauman, Zygmunt. 1982. *Memories of Class: The Pre-History and After-Life of Class*. London: Routledge and Kegan Paul.

Bauman, Zygmunt. 2005. *Work, Consumerism and the New Poor* (2nd edition). Maidenhead: Open University Press.

Baylis, Françoise. 1989. "Persons with Moral Expertise and Moral Experts: Wherein Lies the Difference?" In *Clinical Ethics: Theory and Practice*, edited by Barry Hoffmaster, Benjamin Freedman, and Gwen Fraser, 89–99. Clifton, NJ: Humana Press.

BBC. 2014. "Iran Internet: Hassan Rouhani Tells Clerics Web Is Vital." BBC News, September 1, 2014. http://www.bbc.com/news/world-middle-east-29017729.

Beer, David, 2013. *Popular Culture and New Media: The Politics of Circulation*. Basingstoke: Palgrave Macmillan.

Beer, David. 2009. "Power through the Algorithm? Participatory Web Cultures and the Technological Unconscious." *New Media & Society* 11, no. 6: 985–1002.

Bell, Daniel. 1973. *The Coming of Post-Industrial Society: A Venture in Social Forecasting.* New York: Basic Books.

Belsey, Catherine. 2002. *Post-Structuralism: A Very Short Introduction.* Oxford: Oxford University Press.

Benford, Steve, Chris Greenhalgh, Bob Anderson, Rachel Jacobs, Mike Golembewski, Marina Jirotka, Bernd Carsten Stahl, et al. 2015. "The Ethical Implications of HCI's Turn to the Cultural." *ACM Transactions on Computer-Human Interaction* 22, no. 5: 24:1–24:37.

Beniger, James. 1986. *The Control Revolution: Technological and Economic Origins of the Information Society.* Cambridge, Massachusetts: Harvard University Press.

Bennett, W. Lance. 1990. "Toward a Theory of Press-State Relations in the United States." *Journal of Communication* 40, no. 2: 103–127.

Bennett, W. Lance. 2005. *News: The Politics of Illusion* (6th edition). New York: Longman.

Bentham, Jeremy. 1988. *Introduction to the Principles of Morals and Legislation.* Buffalo: Prometheus Books.

Berdichevsky, Daniel, and Erik Neuenschwander. 1999. "Toward an Ethics of Persuasive Technology." *Communications of the ACM* 42, no. 5: 51–58.

Bernstein, William J. 2013. *Masters of the Word. How Media Shaped History from the Alphabet to the Internet.* London: Atlantic Books.

Bertrand, Marianne, and Sendhil Mullainathan. 2004. "Are Emily and Greg More Employable than Lakisha and Jamal? A Field Experiment on Labor Market Discrimination." *American Economic Review* 94, no. 4: 991–1013.

Bierstedt, Robert. 1950. "An Analysis of Social Power." *American Sociological Review* 15, no. 6: 730–738.

Bijker, Wiebe. 1993. "Do Not Despair: There Is Life after Constructivism." *Science, Technology & Human Values* 18: 113–138.

Bijker, Wiebe. 1995. *Bikes, Bakelite, and Bulbs. Steps Toward a Theory of Socio-Technical Change.* Cambridge: MIT Press.

Bintliff, Barbara. 2005. "Electronic Resources or Print Resources." *Perspectives: Teaching Legal Research and Writing* 14, no. 1: 23–25.

Bloch, Maurice. 1974. "Symbols, Song, Dance and Features of Articulation: Is Religion an Extreme Form of Traditional Authority?." *European Journal of Sociology* 15, no. 1: 54-81.

BMVI. 2017a. Ethics Commission. Automated and Connected Driving. June 2017. https://www.bmvi.de/SharedDocs/EN/publications/report-ethics-commission.pdf.

BMVI. 2017b. Ethics Commission on Automated Driving Presents Report. 28 August, 2017. https://www.bmvi.de/SharedDocs/EN/PressRelease/2017/084-ethic-commission-report-automated-driving.html.

Bok, Sissela. 1979. *Lying.* New York: Vintage Books.

Bolt, L.L.E. 2007. "True to Oneself? Broad and Narrow Ideas on Authenticity in the Enhancement Debate." *Theoretical Medicine and Bioethics* 28, no. 4: 285–300.

Bonnefon, Jean-François, Azim Shariff, and Iyad Rahwan. 2016. "The Social Dilemma of Autonomous Vehicles." *Science* 352, no. 6293: 1573–1576.

Borch, Christian, and Ann-Christina Lange. 2016. "High-Frequency Trader Subjectivity: Emotional Attachment and Discipline in an Era of Algorithms." *Socio-Economic Review* 15, no. 2: 283–306.

Boyce, David E., and Huw C W L Williams. 2015. *Forecasting Urban Travel: Past, Present and Future.* Cheltenham: Edward Elgar.

Bradbury, Jim. 1985. *The Medieval Archer.* New York: St. Martin's Press.

Bradford, Ernle. 1974. *The Sword and Scimitar.* London: Victor Gollancz Ltd.

Brenner, Sydney. 2000. "Genomics—The End of the Beginning." *Science* 287, no. 5461: 2173–2174.

Brey, Philip. 1998. "The Politics of Computer Systems and the Ethics of Design." In *Computer Ethics: Philosophical Enquiry*, edited by Jeroen van den Hoven, 64–75. Rotterdam: Rotterdam University Press.

Brey, Philip. 2000. "Disclosive Computer Ethics", *Computers and Society*, (December 2010), 10–16.

Brey, Philip. 2005. "Artifacts as Social Agents." In *Inside the Politics of Technology: Agency and Normativity in the Co-Production of Technology and Society*, edited by Hans Harbers, 61–84. Amsterdam: Amsterdam University Press.

Brey, Philip. 2008. "The Technological Construction of Social Power." *Social Epistemology* 22, no. 1: 71–95.

Brey, Philip. 2010. "Values in Technology and Disclosive Computer Ethics." In *Cambridge Handbook of Information and Computer Ethics*, edited by Luciano Floridi, 41–58. Cambridge: Cambridge University Press.

Brey, Philip. 2014. "From Moral Agents to Moral Factors: The Structural Ethics Approach." In *The Moral Status of Technical Artefacts*, edited by Peter Kroes and Peter-Paul Verbeek, 125–142, Dordrecht: Springer.

Brey, Philip. 2017. "Ethics of Emerging Technology." In *The Ethics of Technology: Methods and Approaches*, edited by Sven Ove Hansson, 175–191. London: Rowman & Littlefield.

Bridge, Mark. 2018. "Marijuana Growers Can Chill Out and Let Robot Take the Strain." *The Times*, August 13, 2018. https://www.thetimes.co.uk/article/marijuana-growers-can-chill-out-and-let-robot-take-the-strain-b8mjmcgwm.

Bucher, Taina. 2012. "Want to Be on Top? Algorithmic Power and the Threat of Invisibility on Facebook." *New Media and Society* 14, no. 7: 1164–1180.

Bucher, Taina. 2016. "The Algorithmic Imaginary: Exploring the Ordinary Affects of Facebook Algorithms." *Information, Communication & Society* 20, no. 1: 30–44.

Burgess, Kaya. 2018. "AI Will Give Eton education for All, Says Sir Anthony Seldon." *The Times*, October 5, 2018. https://www.thetimes.co.uk/article/ai-will-give-eton-education-for-all-says-sir-anthony-seldon-pwrwb9pz5.

Burrell, Jenna. 2016. "How the Machine 'Thinks': Understanding Opacity in Machine Learning Algorithms." *Big Data & Society* 3, no. 1: 1–12.

Buzan, Barry. 2007. *People, States and Fear*. Colchester, England: ECPR Press.

Callon, Michel, and Bruno Latour. 1992. "Don't Throw the Baby Out with the Bath School! A Reply to Collins and Yearley." In *Science as Practice and Culture*, edited by Andrew Pickering, 343–368. Chicago University of Chicago Press.

Calo, Ryan. 2016. "Privacy, Vulnerability, and Affordance." *DePaul Law Review* 66, 591–602.

Carnap, Rudolf. 1928. *The Logical Structure of the World*. Berkeley: University of California Press.

Castells, Manuel. 2007. "Communication, Power and Counter-Power in the Network Society." *International Journal of Communication* 1, 238–266.

Castells, Manuel. 2009. *Communication Power*. Oxford: Oxford University Press.

Castells, Manuel. 2010a. *The Information Age Economy, Society, and Culture. Volume I: The Rise of the Network Society* (2nd edition, with a new preface). Chichester: Wiley-Blackwell.

Castells, Manuel. 2010b. *The Information Age Economy, Society, and Culture. Volume II: The Power of Identity* (2nd edition, with a new preface). Chichester: Wiley-Blackwell.

Castells, Manuel. 2010c. *The Information Age Economy, Society, and Culture. Volume III: End of Millennium* (2nd edition, with a new preface). Chichester: Wiley-Blackwell.

Charles, Taylor. 1991. *The Ethics of Authenticity*. Cambridge, MA: Harvard University Press.

Cheney-Lippold, John. 2011. "A New Algorithmic Identity: Soft Biopolitics and the Modulation of Control." *Theory, Culture & Society* 28, 6: 164–181.

Choi, Sungho, and Michael Fara. 2018. "Dispositions." *The Stanford Encyclopedia of Philosophy* (Fall 2018), edited by Edward N. Zalta. https://plato.stanford.edu/archives/fall2018/entries/dispositions.

Christiano, Tom. 2018. "Democracy." *The Stanford Encyclopedia of Philosophy*, edited by Edward N. Zalta. https://plato.stanford.edu/archives/fall2018/entries/democracy.

Christman, John. 2018. "Autonomy in Moral and Political Philosophy." *The Stanford Encyclopedia of Philosophy*, edited by Edward N. Zalta. https://plato.stanford.edu/archives/spr2018/entries/autonomy-moral.

Ci, Jiwei. 2011. "Evaluating Agency: A Fundamental Question for Social and Political Philosophy." *Metaphilosophy* 42, no. 3: 262–281.

Clark, Marilyn, and Anna Grech. 2017. *Journalists under Pressure: Unwarranted Interference, Fear and Self-Censorship in Europe*. Strasbourg: Council of Europe Publishing.

Clarke, Adele E., Janet K. Shim, Laura Mamo, Jennifer Ruth Fosket, and Jennifer R. Fishman. 2003. "Biomedicalization: Technoscientific Transformations of Health, Illness, and US Biomedicine." *American Sociological Review* 68, no. 2: 161–194.

Clarkson, John, Roger Coleman, Simeon Keates, and Cherie Lebbon (eds). 2003. *Inclusive design: Design for the whole population.* London: Springer-Verlag.

Clegg, Stewart. 1989. *Frameworks of Power.* London: Sage.

Coeckelbergh, Mark. 2013. *Human Being @ Risk. Enhancement, Technology, and the Evaluation of Vulnerability Transformations.* Dordrecht: Springer.

Coleman, Roger, Simeon Keates, and Cherie Lebbon (Eds.). 2003. *Inclusive Design: Design for the Whole Population.* Heidelberg: Springer-Verlag.

Connolly, William. 1993. *The Terms of Political Discourse* (3rd edition). Princeton, NJ: Princeton University Press.

Connor, Desmond M. 1988. "A New Ladder of Citizen Participation." *National Civic Review* 77, no. 3: 249–257.

Cook, Karen Schweers. 2005. "Networks, Norms, and Trust: The Social Psychology of Social Capital." *Social Psychology Quarterly* 68, no. 1: 4–14.

Cooper, Frank Rudy. 2015. "Always Already Suspect: Revising Vulnerability Theory." *North Carolina Law Review* 93, no. 5: 1339–1379.

Corona, Brezina. 2006. *Al-Khwarizmi: The Inventor of Algebra.* New York: The Rosen Publishing Group.

Crosthwaite, Jan. 1995. "Moral Expertise: A Problem in the Professional Ethics of Professional Ethicists." *Bioethics* 9, no. 4: 361–379.

Currie, Wendy L., and Thomas Lagoarde-Segot. 2017. "Financialization and Information Technology: Themes, Issues and Critical Debates—Part I." *Journal of Information Technology* 32, 211–217.

Dahl, Robert A., and Charles Lindblom. 1953. *Politics, Economics and Welfare.* New York: Harper & Bros.

Dahl, Robert A. 1957. "The Concept of Power." *Behavioral Science* 2, no. 3: 201–215.

Dahl, Robert A. 1961. *Who Governs? Democracy and Power in an American City.* New Haven, CT: Yale University.

Danaher, John. 2016. "The Threat of Algocracy: Reality, Resistance and Accommodation." *Philosophy & Technology* 29, no. 3: 245–268.

Dehghan, Saeed Kamali. 2013a. "Iran's President Signals Softer Line on Web Censorship and Islamic Dress Code." *The Guardian*, July 2, 2013. http://www.theguardian.com/world/2013/jul/02/iran-president-hassan-rouhani-progressive-views.

Dehghan, Saeed Kamali. 2013b. "Iranian Ministers Embrace Social Media Despite Ban." *The Guardian*, September 9, 2013. http://www.theguardian.com/world/2013/sep/09/iranian-ministers-embrace-social-media.

Dehghan, Saeed Kamali. 2013c. "Hassan Rouhani Suggests Online Freedom for Iran in Jack Dorsey Tweet." *The Guardian*, October 2, 2013. http://www.theguardian.com/world/iran-blog/2013/oct/02/iran-president-hassan-rouhani-internet-online-censorship.

de La Grandville, Olivier. 2008. "Economic Growth." In *International Encyclopedia of the Social Sciences*, Volume 2 (2nd edition), edited by William A. Darity Jr., 485–492. New York: Macmillan.

De Vries, Katja. 2010. "Identity, Profiling Algorithms and a World of Ambient Intelligence." *Ethics and Information Technology* 12, no.1: 71–85.

Deleuze, Gilles. 1988. *Foucault.* Minneapolis: University of Minnesota Press.

Devon, Richard, and Ibo Van de Poel. 2004. "Design Ethics: The Social Ethics Paradigm." *International Journal of Engineering Education* 20, no. 3: 461–469.

Dewey, John. 1937 "Democracy and Educational Administration." *School and Society* 45, 457–467.

Diakopoulos, Nicholas. 2014. "Algorithmic-Accountability: The Investigation of Black Boxes." *Tow Center for Digital Journalism*. December 2013. https://academiccommons.columbia.edu/doi/10.7916/D8ZK5TW2.

Dickson, Bruce J. 2008. *Wealth into Power: The Communist Party's Embrace of China's Private Sector.* Cambridge: Cambridge University Press.

Doctorow, Cory. 2008. *Little Brother*. New York: Tor Books.

Doctorow, Cory. 2012. "Censorship Is Inseparable from Surveillance." *The Guardian*, March 2, 2012. https://www.theguardian.com/technology/2012/mar/02/censorship-inseperable-from-surveillance.

Dohrenwend, Robert E. 2012. "The Spear: An Effective Weapon since Antiquity." *Revista de Artes Marciales Asiáticas* 2, no. 2: 8–35.

Dosi, Giovanni, Christopher Freeman, Richard Nelson, Gerald Silverberg, and Luc Soete (Eds.). 1988. *Technical Change and Economic Theory*. London: Pinter Publishers.

Dowd, Nancy E. 2013. "Unfinished Equality: The Case of Black Boys." *Indiana Journal Law & Social Equality* 2, no. 1: 36–61.

Dowding, Keith. 1991. *Rational Choice and Political Power*. Aldershot: Edward Elgar.

Duhaime-Ross, Arielle. 2014. "Apple Promised an Expansive Health App, so Why Can't I Track Menstruation?" *The Verge*. September 25, 2014. https://www.theverge.com/2014/9/25/6844021/apple-promised-an-expansive-health-app-so-why-cant-i-track.

Dunn, John. 1984. "The Concept of 'Trust' in the Politics of John Locke." *Philosophy in History*, edited by Richard Rorty, J. B. Schneewind, and Quentin Skinner, 279–301. Cambridge: Cambridge University Press.

Dunn, John. 1988. "Trust and Political Agency." In *Trust: Making and Breaking Cooperative Relations*, edited by Diego Gambetta, 73–93. Oxford: Blackwell.

Dupuy, Jean-Pierre. 2009. *On the Origins of Cognitive Science: The Mechanization of the Mind*. Cambridge, MA: MIT Press.

Eccles, T. 1981. *Under New Management: The Story of Britain's Largest Worker Cooperative*. London: Pan.

Elliott, Kevin C. 2006. "An Ethics of Expertise Based on Informed Consent." *Science and Engineering Ethics* 12, 637–661.

Emanuel, Kerry. 2005. *Divine Wind: The History and Science of Hurricanes*. Oxford: Oxford University Press.

Epstein, Robert, and Ronald E. Robertson. 2015. "The Search Engine Manipulation Effect (SEME) and Its Possible Impact on the Outcomes of Elections." *Proceedings of the National Academy of Sciences* 112, no. 33: E4512–E4521.

European Commission. 2012. *Responsible Research and Innovation: Europe's Ability to Respond to Societal Challenges*. Luxembourg: Publications Office of the European Union.

Fay, Brian. 1987. *Critical Social Science: Liberation and Its Limits*. Ithaca, NY: Cornell University Press.

Feenberg, Andrew. 1991. *Critical Theory of Technology*. New York: Oxford University Press.

Feenberg, Andrew. 1992. "Subversive Rationalization: Technology, Power, and Democracy." *Inquiry* 35, no. 3–4: 301–322.

Feenberg, Andrew. 1995. *Alternative Modernity: The Technical Turn in Philosophy and Social Theory*. Berkeley: University of California Press.

Feenberg, Andrew. 1999. *Questioning Technology*. London: Routledge.

Feenberg, Andrew. 2002. *Transforming Technology: A Critical Theory Revisited*. Oxford University Press.

Feenberg, Andrew. 2005. "Critical Theory of Technology: An Overview." *Tailoring Biotechnologies* 1, no. 1: 47–64.

Feenberg, Andrew. 2017. *Technosystems: The Social Life of Reason*. Cambridge, MA: Harvard University Press.

Ferguson, Douglas A., and Clark F. Greer. 2016. "Reaching a Moving Target: How Local TV Stations Are Using Digital Tools to Connect with Generation C." *International Journal on Media Management* 18, no. 3–4: 141–161.

Figgis, John Neville. 1914. *The Divine Right of Kings* (2nd edition). Cambridge: Cambridge University Press.

Flanagan, Mary, Daniel C. Howe, and Helen Nissenbaum. 2008. "Embodying Values in Technology: Theory and Practice." In *Information Technology and Moral Philosophy*, edited by Jeroen van den Hoven and John Weckert, 322–353. Cambridge: Cambridge University Press.

Fleck, Ludwik. 1979. *Genesis and Development of a Scientific Fact*. Chicago: Chicago University Press.

Floridi, Luciano. 2008. "The Method of Levels of Abstraction." *Minds and Machines* 18: 303–329.

Floridi, Luciano. 2012. "Big Data and Their Epistemological Challenge." *Philosophy & Technology* 25, no. 4: 435–437.

Floridi, Luciano. 2014. "Artificial Agents and Their Moral Nature." In *The Moral Status of Technical Artefacts*, edited by Peter Kroes, and Peter-Paul Verbeek, 185–212. Dordrecht: Springer.

Flyvbjerg, Bent. 1998. *Rationality and Power: Democracy in Practice*. Chicago, IL: Chicago University Press.

Fogg, Brian J. 2003. *Persuasive Technology: Using Computers to Change What We Think and Do*. San Francisco, CA: Morgan Kaufmann.

Foucault, Michel. 1976. *La Volonté de Savoir*. Paris: Editions Gallimard.

Foucault, Michel. 1977. *Discipline and Punish: The Birth of the Prison*. Translated by Alan Sheridan. New York: Vintage Books.

Foucault, Michel. 1978. *The History of Sexuality, Volume 1: An Introduction*. Translated by Robert Hurley. New York: Pantheon Books.

Foucault, Michel. 1980. *Power/Knowledge: Selected Interviews and Other Writings 1972–1977*, edited by Colin Gordon. New York: Pantheon Books.

Foucault, Michel. 1982. "The Subject and Power." *Critical Inquiry* 8, no. 4: 777–795.

Foucault, Michel. 1997. "Sexuality and Solitude." In *Ethics: Subjectivity and Truth: The Essential Works of Michel Foucault 1954–1984, Volume 1*, edited by Paul Rabinow, 175–184. New York: The New Press.

Foucault, Michel. 2003. *Society Must Be Defended: Lectures at the Collège de France, 1975–1976*. Translated by David Macey. New York: Picador.

Fournier, Tom. 2016. "Will My Next Car Be a Libertarian or a Utilitarian?: Who Will Decide?" *IEEE Technology and Society Magazine* 35, no. 2: 40–45.

Frauenberger, Christopher, Marjo Rauhala, and Geraldine Fitzpatrick. 2017. "In-Action Ethics." *Interacting with Computers* 29, no. 2: 220–236.

French, John R. P., and Bertram Raven. 1959. "The Bases of Social Power." In *Studies in Social Power*, edited by Dorwin Cartwright, 259–269. Ann Arbor, MI: Institute for Social Research.

Friedman, Batya (Ed). 1997. *Human Values and the Design of Computer Technology*. Cambridge: Cambridge University Press.

Friedman, Batya, and Helen Nissenbaum. 1996. "Bias in Computer Systems." *ACM Transactions on Information Systems* 14, no. 3: 330–347.

Friedman, Batya, Peter H. Kahn Jr., and Alan Borning. 2006. "Value Sensitive Design and Information Systems." In *Human-Computer Interaction in Management Information Systems: Foundations*, edited by Ping Zhang and Dennis F. Galletta, 348–372. New York: M.E. Sharpe.

Friedman, Batya, Peter H. Kahn Jr., and Alan Borning. 2008. "Value Sensitive Design and Information Systems." In *The Handbook of Information and Computer Ethics*, edited by Kenneth Einar Himma and Herman T. Tavani, 69–101. Hoboken, NJ: John Wiley & Sons.

Fuchs, Christian. 2008. *Internet and Society: Social Theory in the Information Age*. New York: Routledge.

Fung, Archon. 2006. "Varieties of Participation in Complex Governance." Public Administration Review 66, 66–75.

Gale, W. K. V. 1973. "The Bessemer Steelmaking Process." *Transactions of the Newcomen Society* 46, no. 1: 17–26.

Gallie, Walter Bryce. 1956. "Essentially Contested Concepts." *Proceedings of the Aristotelian Society* 56: 167–198.

Galperin, Hernan. 2004. *New Television: Old Politics*. Cambridge: Cambridge University Press.

Garsten, Bryan. 2009. *Saving Persuasion: A Defense of Rhetoric and Judgment*. Cambridge, MA: Harvard University Press.

Geertz, Clifford. 1973. *The Interpretation of Cultures*. New York: Basic Books.

Gellner, Ernest. 1979. "The Withering Away of the Dentistry State." *Review (Fernand Braudel Center)* 2, no. 3: 461–472.

Gerschenkron, Alexander. 1943. *Bread and Democracy in Germany*. Berkeley: University of California Press.

Gibson, Quentin. 1971. "Power." *Philosophy of Social Sciences* 1: 101–112.

Giddens, Anthony. 1981. *A Contemporary Critique of Historical Materialism. Volume 1: Power, Property and the State*. Berkeley: University of California Press.

Giddens, Anthony. 1984. *The Constitution of Society. Outline of the Theory of Structuration*. Cambridge: Polity Press.

Giddens, Anthony. 1985a. *A Contemporary Critique of Historical Materialism. Volume 2: The Nation-State and Violence*. Cambridge: Polity Press.

Giddens, Anthony. 1985b. "Marx's Correct Views on Everything." *Theory and Society* 14, no. 2: 167–174.

Giddens, Anthony. 1991. *Consequences of Modernity*. Oxford: Polity Press.

Gisler, Galen R. 2008. "Tsunami Simulations." *Annual Review of Fluid Mechanics* 40, 71–90.

Godwin, William. 2009 (1794). *Caleb Williams*, edited by Pamela Clemit. Oxford: Oxford University Press.

Gogoll, Jan, and Julian F. Müller. 2017. "Autonomous Cars: In Favor of a Mandatory Ethics Setting." *Science and Engineering Ethics* 23, no. 3: 681–700.

Goldman, Alvin I. 1972. "Toward a Theory of Social Power." *Philosophical Studies* 23, 221–268.

Goldman, Eric. 2006. "Search Engine Bias and the Demise of Search Engine Utopianism." *Yale Journal of Law and Technology* 8, 1: 188–200.

Goldsmith, Jack, and Tim Wu. 2006. *Who Controls the Internet? Illusions of Borderless World*. Oxford: Oxford University Press.

Golumbia, David. 2009. *The Cultural Logic of Computation*. Cambridge: Harvard University Press.

Goodall, Noah J. 2016. "Can You Program Ethics into a Self-Driving Car?" *IEEE Spectrum* 53, no. 6: 28–58.

Goodin, Robert E. 1995. *Utilitarianism as a Public Philosophy*. Cambridge: Cambridge University Press.

Goodman, Bryce, and Seth Flaxman. 2017. "European Union Regulations on Algorithmic Decision-Making and a 'Right to Explanation'." *AI Magazine* 38, no. 3: 50–57.

Goodman, Nelson. 1954. *Fact, Fiction and Forecast*. Cambridge, MA: Harvard University Press.

Gould, Carol C. 1988. *Rethinking Democracy: Freedom and Social Cooperation in Politics, Economics and Society*. Cambridge: Cambridge University Press.

Gramsci, Antonio. 1971. *Selections from the Prison Notebooks*. New York: International Publishers.

Gray, Colin S. 1975. "New Weapons and the Resort to Force." *International Journal* 30, no. 2: 238–258.

Gray, John. 1978. "On Liberty, Liberalism and Essential Contestability." *British Journal of Political Science* 8, no. 4: 385–402.

Greene, Joshua D. 2016. "Our Driverless Dilemma." *Science* 352, no. 6293: 1514–1515.

Griffin, James. 1997. "Incommensurability: What's the Problem?." In *Incommensurability, Incomparability and Practical Reason*, edited by Ruth Chang, 35–51. Cambridge: Harvard University Press.

Grossman, Gene M., and Elhanan Helpman. 1991. *Innovation and Growth in the Global Economy*. Cambridge, Massachusetts: MIT Press.

Grossman, Gene, and Elhanan Helpman. 1994. "Protection for Sale." *American Economic Review* 84, no. 4: 833–850.

Guignon, Charles B. 1993. "Authenticity, Moral Values, and Psychotherapy." In *The Cambridge Companion to Heidegger*, edited by Charles B. Guignon, 215–239. Cambridge: Cambridge University Press

Gurzawska, Agata, Markus Mäkinen, and Philip Brey. 2017. "Implementation of Responsible Research and Innovation (RRI) Practices in Industry: Providing the Right Incentives." *Sustainability* 9, DOI:10.3390/su9101759.

Gutmann, Amy, and Dennis Thompson. 2004. *Why Deliberative Democracy?*. Princeton, NJ: Princeton University Press.

Habermas, Jürgen. 1996. *Between Facts and Norms: Contributions to a Discourse Theory of Law and Democracy*. Translated by William Rehg. Cambridge: Polity Press.

Hacking, Ian. 1981. "The Archaeology of Foucault." Review of *Power/Knowledge: Selected Interviews and Other Writings, 1972–1977*, by Michel Foucault. *New York Review of Books*, May 14, 1981.

Hajian, Sara, Francesco Bonchi, and Carlos Castillo. 2016. "Algorithmic Bias: From Discrimination Discovery to Fairness-Aware Data Mining." In *Proceedings of the 22nd ACM SIGKDD International Conference on Knowledge Discovery and Data Mining*, 2125–2126.

Halperin, Morton H., Joseph T. Siegle, and Michael M. Weinstein. 2005. *The Democracy Advantage: How Democracies Promote Prosperity and Peace*. New York: Routledge.

Hansson, Sven Ove. 2017. "The Ethics of Doing Ethics of Technology." In *The Ethics of Technology: Methods and Approaches*, edited by Sven Ove Hansson, 239–250. London: Rowman & Littlefield.

Hardin, Russell. 1991. "Trusting Persons, Trusting Institutions." In *Strategy and Choice*, edited by Richard J. Zeckhauser, 185–210. Cambridge, MA: The MIT Press.

Hardin, Russell. 2002. *Trust and Trustworthiness*. New York: The Russell Sage Foundation.

Hardt, Michael, and Antonio Negri. 2000. *Empire*. Cambridge, MA: Harvard University Press.

Harvey, David. 1989. *The Condition of Postmodernity: An Enquiry into the Origins of Cultural Change*. Oxford: Basil Blackwell.

Harvey, Olivia. 2008. Review of *The Politics of Life Itself: Biomedicine, Power, and Subjectivity in the Twenty-First Century*, by Nicolas Rose. *Sociology of Health & Illness* 30, no. 7: 1117–1118.

Haugaard, Mark, and Stuart Clegg. 2009. "Why Power Is the Central Concept of the Social Sciences." in *The Sage Handbook of Power*, edited by Mark Haugaard, and Stuart Clegg, 1–24. London: Sage.

Haugaard, Mark. 1997. *The Constitution of Power: A Theoretical Analysis of Power, Knowledge and Structure*. Manchester: Manchester University Press.

Haugaard, Mark. 2010. "Power: A 'Family Resemblance' Concept." *European Journal of Cultural Studies* 13, no. 4: 419–438.

Haugaard, Mark. 2012. "Editorial: Reflections upon Power over, Power to, Power with, and the Four Dimensions of Power." *Journal of Political Power* 5, no. 3: 353–358. New Haven, CT: Yale University.

Haybron, Daniel M. 2008. "Happiness, the Self and Human Flourishing." *Utilitas* 20, no. 1: 21–49.

Henrich, Joseph, Robert Boyd, Samuel Bowles, Colin Camerer, Ernst Fehr, Herbert Gintis, and Richard McElreath. 2001. "In Search of Homo Economicus: Behavioral Experiments in 15 Small-Scale Societies." *American Economic Review* 91, no. 2: 73–78.

Herman, Edward S., and Noam Chomsky. 1988. *Manufacturing Consent. The Political Economy of the Mass Media*. New York: Pantheon Books.

Hern, Alex. 2016. "Microsoft Scrambles to Limit PR Damage over Abusive AI Bot Tay." *The Guardian*, March 24, 2016. https://www.theguardian.com/technology/2016/mar/24/microsoft-scrambles-limit-pr-damage-over-abusive-ai-bot-tay.

Hessler, Pierre. 2014. "Technology and Economic Power." *Philosophy & Technology* 27, no. 2: 279–283.

Hill, Robin K. 2015. "What an Algorithm Is." *Philosophy & Technology* 29, 1: 35–59.

Hindess, Barry. 1996. *Discourses of Power from Hobbes to Foucault*. Oxford: Blackwell.

Hobbes, Thomas. 1839 (1651). *Leviathan: The English Works of Thomas Hobbes, Volume III*, edited by Sir William Molesworth. London: John Bohn.

Holmes, Helen. 2018. "This Ugly Painting Made by a Robot Just Sold for $432,500." *Observer*, October 25, 2018. https://observer.com/2018/10/ai-created-portrait-of-edmond-belamy-christies-worth-it.

Honoré, Anthony M. 1964. "Can and Can't." *Mind* 73, no. 292: 463–479.

Howard, Philip N., Sheetal D. Agarwal, and Muzammil M. Hussain. 2011. "The Dictators' Digital Dilemma: When Do States Disconnect Their Digital Networks?" *Issues in Technology Innovation* 13, 1–11.

Hsu, Jeremy. 2014. "Iran Eases Restrictions on High-Speed Mobile Internet." *IEEE Spectrum*, September 5, 2014. http://spectrum.ieee.org/tech-talk/telecom/internet/iran-eases-restrictions-on-highspeed-mobile-internet.

Hughes, Thomas Parke. 1983. *Networks of Power Electrification in Western Society, 1880–1930*. Baltimore: The John Hopkins Press.

Hume, David. 1888. *A Treatise of Human Nature*, edited by Lewis Amherst Selby-Bigge. Oxford: Clarendon Press.

Hurst, Greg. 2018. "Cash-Strapped Councils Turn to Algorithms to Spot Children at Risk." *The Times*, September 17, 2018. https://www.thetimes.co.uk/article/cash-strapped-councils-turn-to-algorithms-to-spot-children-at-risk-z7w5b9xb7.

Hyun, Insoo. 2001. "Authentic Values and Individual Autonomy." *The Journal of Value Inquiry* 35, no. 2: 195–208.

Iftode, Cristian. 2019. "Assessing Enhancement Technologies: Authenticity as a Social Virtue and Experiment." *The New Bioethics* (2019): 1–15.

Illich, Ivan. 1973. *Tools for Conviviality*. New York: Harper and Row.

Inglehart, Ronald, and Christian Welzel. 2005. *Modernization, Cultural Change, and Democracy: The Human Development Sequence*. Cambridge University Press, 2005.

Joerges, Bernward. 1988. "Large Technical Systems: Concepts and Issues." In *Development of Large Technical Systems*, edited by Renate Mayntz and Thomas P. Hughes, 9–36. Boulder, CO: Westview Press.

Johnson, David R., and David Post. 1995. "Law and Borders: The Rise of Law in Cyberspace." *Stanford Law Review* 48, no. 5: 1367–1402.

Jonas, Hans. 1974. *Philosophical Essays: From Ancient Creed to Technological Man*. Englewood Cliffs, NJ: Prentice-Hall.

Jonas, Hans. 1984. *The Imperative of Responsibility: In Search of an Ethics for the Technological Age*. Chicago: University of Chicago Press.

Jukes, Stephen, Katy McDonald, and Guy Starkey. 2018. *Understanding Broadcast Journalism*. London: Routledge.

Juth, Niklas. 2011. "Enhancement, Autonomy, and Authenticity." In *Enhancing Human Capacities*, edited by Julian Savulescu, Ruud ter Meulen, and Guy Kahane, 34–48, Oxford: Blackwell.

Kaufmann, Daniel, Aart Kraay, and Massimo Mastruzzi. 2011. "The Worldwide Governance Indicators: Methodology and Analytical Issues." *Hague Journal on the Rule of Law* 3, no. 2: 220–246.

Kelly, Mark G. E. 2009. *The Political Philosophy of Michel Foucault*. London: Routledge.

Kelly, Mark G. E. 2004. "Racism, Nationalism and Biopolitics: Foucault's Society Must Be Defended, 2003." *contretemps*, no. 4.

Kennedy, Emmet. 1979. "'Ideology' from Destutt De Tracy to Marx." *Journal of the History of Ideas* 40, no. 3: 353–368.

Kenny, Anthony. 1975. *Will, Freedom and Power*. Oxford: Blackwell.

Kenny, Anthony. 2006. *A New History of Western Philosophy, Volume III: The Rise of Modern Philosophy*. Oxford: Clarendon Press.

Kifer, Yona, Daniel Heller, Wei Qi Elaine Perunovic, and Adam D. Galinsky. 2013 "The Good Life of the Powerful: The Experience of Power and Authenticity Enhances Subjective Well-Being." *Psychological Science* 24, no. 3: 280–288.

Kitchin, Rob. 2016. "Thinking Critically about and Researching Algorithms." *Information, Communication & Society* 20, no. 1: 14–29.

Knobel, Cory, and Geoffrey C. Bowker. 2011. "Computing Ethics—Values in Design." *Communications of the ACM* 54, no. 7: 26–26.

Knorr Cetina, Karin D., and Urs Bruegger. 2002. "Traders Engagement with Markets: A Postsocial Relationship." *Theory, Culture and Society* 19, no. 5/6: 161–185.

Koops, Bert-Jaap, Ilse Oosterlaken, Henny Romijn, Tsjalling Swierstra, and Jeroen van den Hoven. 2015. *Responsible Innovation 2: Concepts, Approaches, and Applications*. Heidelberg: Springer.

Kovács, József. 2010. "The Transformation of (Bio) Ethics Expertise in a World of Ethical Pluralism." *Journal of Medical Ethics* 36, no. 12: 767–770.

Kozinets, Robert V. 2008. "Technology/Ideology: How Ideological Fields Influence Consumers' Technology Narratives." *Journal of Consumer Research* 34, no. 6: 865–881.

Kraemer, Felicitas, Kees Van Overveld, and Martin Peterson. 2011. "Is There an Ethics of Algorithms?" *Ethics and Information Technology* 13, no. 3: 251–260.

Kramer, Adam D. I., Jamie E. Guillory, and Jeffrey T. Hancock. 2014. "Experimental Evidence of Massive-Scale Emotional Contagion through Social Networks." *Proceedings of the National Academy of Sciences* 111, no. 24: 8788–8790.

Kraut, Richard. 2018. "Aristotle's Ethics." *The Stanford Encyclopedia of Philosophy* (Summer 2018), edited by Edward N. Zalta. https://plato.stanford.edu/archives/sum2018/entries/aristotle-ethics.

Krusell, Per, and Jose-Victor Rios-Rull. 1996. "Vested Interests in a Positive Theory of Growth and Stagnation." *The Review of Economic Studies* 63, no. 2: 301–329.

Kuznets, Simon. 1968. *Towards a Theory of Economic Growth*. New Haven, CT: Yale University Press.

Lacewing, Michael. 2015. *Philosophy for A2: Ethics and Philosophy of Mind*. London: Routledge.

Laclau, Ernesto, and Chantal Mouffe. 1985. *Hegemony and Socialist Strategy*. London: Verso.

Ladd, John. 1991. "Bhopal: An Essay on Moral Responsibility and Civic Virtue." *Journal of Social Philosophy* 32, 73–91.

Laird, Frank N. 1993. "Participatory Analysis, Democracy, and Technological Decision Making." *Science, Technology, & Human Values* 18, no. 3: 341–361.

Lash, Scott. 2007. "Power After Hegemony: Cultural Studies in Mutation?" *Theory, Culture & Society* 24, no. 3: 55–78.

Laslett, Peter. 1988. "Introduction." In *John Locke. Two Treatises of Government*, edited by Peter Laslett, 3–126. Cambridge: Cambridge University Press.

Latour, Bruno. 1991. "Technology Is Society Made Durable." In *A Sociology of Monsters: Essays on Power, Technology and Domination*, edited by John Law, 103–131. London: Routledge.

Latour, Bruno. 1992. "Where Are the Missing Masses? The Sociology of a Few Mundane Artifacts." In *Shaping Technology/Building Society: Studies in Sociotechnical Change*, edited by Wiebe E. Bijker, and John Law, 225–258. Cambridge, MA: MIT Press.

Lattimore, Owen. 1962. *Studies in Frontier History*. London: Oxford University Press.

Law, John. 2012. "Technology and Heterogeneous Engineering: The Case of Portuguese Expansion." In *The Social Construction of Technological Systems: New Directions in the Sociology and History of Technology* (anniversary edition), edited by Wiebe E. Bijker, Thomas P. Hughes, and Trevor Pinch, 105–127. Cambridge, MA: MIT Press.

Lay, Kat. 2018. "Algorithm Rivals Doctors in Lung Disease Diagnosis." *The Times*, October 22, 2018. https://www.thetimes.co.uk/article/algorithm-rivals-doctors-in-lung-disease-diagnosis-8cz03xzng.

Layer, David H. 2001. "Digital Radio Takes to the Road." *Spectrum IEEE* 38, no. 7: 40–46.

Lebow, Richard Ned. 2010. *Forbidden Fruit: Counterfactuals and International Relations*. Princeton, NJ: Princeton University Press.

Lessig, Lawrence. 2001. *The Future of Ideas: The Fate of the Commons in a Connected World*. New York: Random House.

Lessig, Lawrence. 2006. *Code*. Version 2.0. New York: Basic Books.

Levitas, Ruth. 2005. *The Inclusive Society? Social Exclusion and New Labour* (2nd edition). Basingstoke: Palgrave Macmillan.

Lewis, Tanya. 2014. "Apple's Health App Tracks Almost Everything, Except Periods." *Livescience*, September 26, 2014. https://www.livescience.com/48040-apple-healthkit-lacks-period-tracker.html.

Littauer, Mary Aiken. 1972. "The Military Use of the Chariot in the Aegean in the Late Bronze Age." *American Journal of Archaeology* 76, no. 2: 145–157.

Locke, John. 1854 (1689). *The Works of John Locke. Volume I: Philosophical Works*. London: Henry G. Bohn.

Locke, John. 1988. *John Locke: Two Treatises of Government* (Student edition), edited by Peter Laslett. Cambridge: Cambridge University Press.

Lord Acton, John Emerich Edward Dalberg. 1887. *Acton-Creighton Correspondence*. https://oll.libertyfund.org/titles/acton-acton-creighton-correspondence.

Luhmann, Niklas. 2017. *Trust and Power*. Cambridge: Polity Press.

Lukes, Steven. 1974. *Power: A Radical View*. London: Palgrave Macmillan.

Lukes, Steven. 2005. *Power: A Radical View* (2nd expanded edition). London: Palgrave Macmillan.

Lukes, Steven. 2007. "Power and the Battle for Hearts and Minds: On the Bluntness of Soft Power." In *Power in World Politics*, edited by Felix Berenskoetter and M. J. Williams, 83–97. London: Routledge.

MacIntyre, Alasdair. 1972. "Is a Science of Comparative Politics Possible?" In *Philosophy, Politics and Society*, edited by Peter Laslett, Walter Garrison Runciman, and Quentin Skinner, 8–26, Oxford: Basil Blackwell.

Mackay, Hughie, and Gareth Gillespie. 1992. "Extending the Social Shaping of Technology Approach: Ideology and Appropriation." *Social Studies of Science* 22, no. 4: 685–716.

Mackenzie, Catriona, and Mary Walker. 2015. "Neurotechnologies, Personal Identity, and the Ethics of Authenticity." In *Handbook of Neuroethics*, edited by Jens Clausen and Neil Levy, 373–392. Dordrecht: Springer.

MacKenzie, Donald. 2015. "Mechanizing the Merc: The Chicago Mercantile Exchange and the Rise of High-Frequency Trading." *Technology and Culture*, 56, 646–675.

MacKenzie, Donald. 2018. "Material Signals: A Historical Sociology of High-Frequency Trading." *American Journal of Sociology* 123, no. 6: 1635–1683.

Mackie, John L. 1977. "Dispositions, Grounds, and Causes." *Synthese* 34, no. 4: 361–369.

Mager, Astrid. 2012. "Algorithmic Ideology: How Capitalist Society Shapes Search Engines." *Information, Communication & Society* 15, no. 5: 769–787.

Maier, John. 2018. "Abilities." *The Stanford Encyclopedia of Philosophy* (Spring 2018), edited by Edward N. Zalta. https://plato.stanford.edu/archives/spr2018/entries/abilities.

Mali, Franc, Toni Pustovrh, Blanka Groboljsek, and Christopher Coenen. 2012. "National Ethics Advisory Bodies in the Emerging Landscape of Responsible Research and Innovation." *Nanoethics* 6, no. 3: 167–184.

Mannheim, Karl. 1936. *Ideology and Utopia: An Introduction to the Sociology of Knowledge*. London: Routledge.

Mann, Gideon, and Cathy O'Neil. 2016. "Hiring Algorithms Are Not Neutral." *Harvard Business Review*, December 9, 2016. https://hbr.org/2016/12/hiring-algorithms-are-not-neutral.

Mann, Michael. 1986. *The Sources of Social Power, Volume I: A History of Power from the Beginning to A.D. 1760*. Cambridge: Cambridge University Press.

Mann, Michael. 1993. *The Sources of Social Power, Volume II: The Rise of Classes and Nation-States, 1760–1914*. Cambridge: Cambridge University Press.

Mann, Michael. 2012. *The Sources of Social Power, Volume 3: Global Empires and Revolution, 1890–1945*. Cambridge: Cambridge University Press.

Mann, Michael. 2013. *The Sources of Social Power, Volume 4: Globalizations, 1945–2011*. Cambridge: Cambridge University Press.

Maurer, Markus, J. Christian Gerdes, Barbara Lenz, and Hermann Winner. 2015. *Autonomous Driving*. Heidelberg: Springer.

McCall, Rod, and Lynne Baillie. 2017. "Ethics, Privacy, and Trust in Serious Games." In *Handbook of Digital Games and Entertainment Technologies*, edited by Nakatsu, Ryohei, Matthias Rauterberg, and Paolo Ciancarini, 611–640, Singapore: Springer.

McFee, Graham. 2015. *How to Do Philosophy: A Wittgensteinian Reading of Wittgenstein*. Cambridge: Cambridge Scholars Publishing.

Mead, Walter Russell. 2005. *Power, Terror, Peace, and War: America's Grand Strategy in a World at Risk*. New York: Vintage Books.

Mele, Alfred R. 2003. "Agents' Abilities." *Noûs* 37, no. 3: 447–470.

Mellor, David Hugh. 1974. "In Defence of Dispositions." *The Philosophical Review* 83, 157–181.

Michelfelder, Diane P., Galit Wellner, and Heather Wiltse. 2017. "Designing Differently: Toward a Methodology for an Ethics." In *The Ethics of Technology: Methods and Approaches*, edited by Sven Ove Hansson, 193–218. London: Rowman & Littlefield.

Mill, John Stuart. 1991. *Considerations on Representative Government*. Buffalo, NY: Prometheus Books.

Mill, John Stuart. 1999 (1859). *On Liberty*, edited by Edward Alexander. Peterborough, Ontario: Broadview Press.

Mitcham, Carl. 2005. "Values and Valuing." In *Encyclopedia of Science, Technology, and Ethics*, edited by Carl Mitcham, 4. Detroit, MI: Macmillan Reference.

Mitchell, Brian R. 1993. *International Historical Statistics*. New York: Stockton.

Mittelstadt, Brent Daniel, and Luciano Floridi. 2016 "The Ethics of Big Data: Current and Foreseeable Issues in Biomedical Contexts." *Science and Engineering Ethics* 22, no. 2: 303–341.

Mittelstadt, Brent Daniel, Patrick Allo, Mariarosaria Taddeo, Sandra Wachter, and Luciano Floridi. 2016. "The Ethics of Algorithms: Mapping the Debate." *Big Data & Society* 3, no. 2: 1–21.

Mokyr, Joel. 1990. *The Lever of Riches: Technological Creativity and Economic Progress*. Oxford: Oxford University Press.

Monro, D. H. 1953. *Godwin's Moral Philosophy*. Oxford: Oxford University Press.

Montesquieu. 1989. *The Spirit of the Laws*. Translated and edited by Anne M. Cohler, Basia Carolyn Miller, and Harold Samuel Stone. Cambridge: Cambridge University Press.

Moran, Terence. P. 2010. *Introduction to the History of Communication: Evolutions and Revolutions*. New York: Peter Lang Publishing.

Morozov, Evgeny. 2011. *The Net Delusion: The Dark Side of Internet Freedom*. New York: PublicAffairs.

Morozov, Evgeny. 2015. "Silicon Valley Exploits Time and Space to Extend the Frontiers of Capitalism." *The Guardian*, November 29, 2015. https://www.theguardian.com/commentisfree/2015/nov/29/silicon-valley-exploits-space-evgeny-morozov.

Morozov, Evgeny. 2016. "Silicon Valley Was Going to Disrupt Capitalism. Now It's Just Enhancing It." *The Guardian*, August 7, 2016. https://www.theguardian.com/commentisfree/2016/aug/07/silicon-valley-health-finance.

Morrison, John, ed. 1995. *The Age of the Galley: Mediterranean Oared Vessels since Pre-Classical Times*. London: Conway Maritime Press.

Morriss, Peter. 2002. *Power: A Philosophical Analysis*. Manchester: Manchester University Press.

Mosse, Werner Eugen. 1992. *An Economic History of Russia, 1856–1914*. London: Taurus.

Motwani, Rajeev, and Prabhakar Raghavan. 1995. *Randomized Algorithms*. Cambridge: Cambridge University Press.

Mumford, Lewis. 1964. "Authoritarian and Democratic Technics." *Technology and Culture* 5: 1–8.

Myers, Fred R. 2001. "Introduction: The Empire of Things." In *The Empire of Things: Regimes of Value and Material Culture*, edited by Fred R. Myers, 3–64. Oxford: James Currey.

Natarajan, Priyamvada. 2014. "What Scientists Really Do." *The New York Review of Books* LXI, no. 16: 64–66.

Nelson, Richard R. 2008. "What Enables Rapid Economic Progress: What are the Needed Institutions?." *Research Policy* 37: 1-11.

Neustadt, Richard. 1991. *Presidential Power and the Modern Presidents*. New York: The Free Press.

Nevelow Mart, Susan . 2017. "The Algorithm as a Human Artifact: Implications for Legal [Re]Search." *Law Library Journal* 109, 3: 387–422.

Newman, Lily Hay. 2015. "Apple Notices That Basically Half the Population Menstruates." *Slate*, June 8, 2015. https://slate.com/technology/2015/06/apple-finally-adds-reproductive-health-tracking-in-healthkit.html.

Neyland, Daniel, and Norma Möllers. "Algorithmic IF . . . THEN Rules and the Conditions and Consequences of Power." *Information, Communication & Society* 20, no. 1: 45–62.

Nickel, Philip, Maarten Franssen, and Peter Kroes. 2010. "Can We Make Sense of the Notion of Trustworthy Technology?" *Knowledge, Technology & Policy* 23, no. 3–4: 429–444.

Nickel, Philip. 2011. "Ethics in E-Trust and E-Trustworthiness: The Case of Direct Computer-Patient Interfaces." *Ethics & Information Technology* 13, 355–363.

Nickel, Philip. 2013. "Trust in Technological Systems." In *Norms in Technology, Philosophy of Engineering and Technology*, edited by Marc J. de Vries, 223–237. Dordrecht: Springer.

Nickel, Philip. 2015. "Design for The Value of Trust." In *Handbook of Ethics, Values, and Technological Design*, edited by Jeroen van den Hoven, Pieter E. Vermaas, and Ibo van de Poel, 551–567. Dordrecht: Springer.

Nihlén Fahlquist, Jessica. 2006. "Responsibility Ascriptions and Vision Zero." *Accident Analysis and Prevention* 38, 1113–1118.

Nihlén Fahlquist, Jessica. 2017. "Responsibility Analysis." In *The Ethics of Technology: Methods and Approaches*, edited by Sven Ove Hansson, 129–142. London: Rowman & Littlefield.

Niker, Fay, and Laura Specker Sullivan. 2018. "Trusting Relationships and the Ethics of Interpersonal Action." *International Journal of Philosophical Studies* 26, no. 2: 173–186.

Nissenbaum, Helen. 2005. "Values in Technical Design." In *Encyclopedia of Science, Technology, and Ethics*, edited by Carl Mitcham, 1xvi–1xx. Detroit, MI: Macmillan Reference.

Nizam Al-Mulk. 2002. *The Book of Government or Rules for Kings: The Siyar Al-Muluk or Siyasat-nama of Nizam Al-Mulk*. Translated by Hubert Darke. London: Routledge.

Noble, Cheryl N., Peter Singer, Jerry Avorn, Daniel Wikler, and Tom L. Beauchamp. 1982. "Ethics and Experts." *The Hastings Center Report* 12, no. 3: 7–15.

Noble, David. 1984. *Forces of Production: A Social History of Industrial Automation*. New York: Knopf.

Noble, Safiya Umoja. 2018. *Algorithms of Oppression: How Search Engines Reinforce Racism*. New York: New York University Press.

Nocetti, Julien. 2015. "Russia's 'Dictatorship-of-the-Law' Approach to Internet Policy." *Internet Policy Review* 4, no. 4, DOI: 10.14763/2015.4.380.

Norman, Donald A. 1988. *The Psychology of Everyday Things*. New York: Basic Books.

North, Douglass C. 1981. *Structure and Change in Economic History*. New York: Norton.

North, Douglass C. 1990. *Institutions, Institutional Change, and Economic Performance*. Cambridge: Cambridge University Press.

North, Douglass C. 1999. *Understanding the Process of Economic Change*. London: Institute of Economic Affairs.

Nye, Joseph S. Jr. 2011. *The Future of Power*. New York: Public Affairs.

Nyholm, Sven, and Jilles Smids. 2016. "The Ethics of Accident-Algorithms for Self-Driving Cars: An Applied Trolley Problem?" *Ethical Theory and Moral Practice* 19, no. 5: 1275–1289.

O'Farrell, Clare. 1982. "Foucault and the Foucauldians", *Economy and Society* 11, no. 4: 449–459.

Oftedal, Gry. 2014. "The Role of Philosophy of Science in Responsible Research and Innovation (RRI): The Case of Nanomedicine." *Life Sciences, Society and Policy*, 10, no. 5. doi: https://doi.org/10.1186/s40504-014-0005-8.

Okoniewski, Elissa A. 2002. "Yahoo!, Inc. v. LICRA: The French Challenge to Free Expression on the Internet." *American University International Law Review* 18, no. 1: 295–339.

O'Neil, Cathy. 2016. *Weapons of Math Destruction: How Big Data Increases Inequality and Threatens Democracy*. New York: Crown.

Oppenheim. Felix E. 1961. *Dimensions of Freedom: An Analysis*. New York: St. Martin's Press.

Östlund, Britt. 2015. "The Benefits of Involving Older People in the Design Process." In *International Conference on Human Aspects of IT for the Aged Population*, 3–14. Cham: Springer.

Owen, Richard, Jack Stilgoe, Phil Macnaghten, Mike Gorman, Erik Fisher, and Dave Guston. 2013. "A Framework for Responsible Innovation." In *Responsible Innovation: Managing the Responsible Emergence of Science and Innovation in Society*, edited by Owen, Richard, John R. Bessant, and Maggy Heintz, 27–50. London: John Wiley & Sons.

Oxford English Dictionary. 2010. *Algorithm*. Accessed online on 2016. https://www.oed.com.

Pak, Richard, Nicole Fink, Margaux Price, Brock Bass, and Lindsay Sturre. 2012. "Decision Support Aids with Anthropomorphic Characteristics Influence Trust and Performance in Younger and Older Adults." *Ergonomics* 55, no. 9: 1059–1072.

Parens, Erik. 2005. "Authenticity and Ambivalence: Toward Understanding the Enhancement Debate." *Hastings Center Report* 35, no. 3: 34–41.

Parsons, Talcott. 1958. "Durkheim's Contribution to the Theory of the Integration of Social Systems." In *Sociological Theory and Modern Society*, edited by Talcott Parsons, 3–34, New York: The Free Press.

Parsons, Talcott. 1963. "On the Concept of Political Power." *Proceedings of the American Philosophical Society* 107, no. 3: 232–262.

Pasquale, Frank. 2015. *The Black Box Society: The Secret Algorithms That Control Money and Information*. Cambridge, MA: Harvard University Press.

Payne, Sebastian. 2014. "Ukrainian President Poroshenko asks Congress for Military Support." *The Washington Post*, September 18, 2014. http://www.washingtonpost.com/news/post-politics/wp/2014/09/18/ukrainian-president-poroshenko-asks-congress-for-military-support.

Petryna, Adriana. 2002. *Life Exposed: Biological Citizens after Chernobyl*. Princeton, NJ: Princeton University Press.

Pfaffenberger, Bryan. 1990. "The Harsh Facts of Hydraulics: Technology and Society in Sri Lanka's Colonization Schemes." *Technology and Culture* 31, no. 3: 361–397.

Pinch, Trevor, and Wiebe Bijker. 2012. "The Social Construction of Facts and Artifacts: Or How the Sociology of Science and the Sociology of Technology Might Benefit Each Other." In *The Social Construction of Technological Systems: New Directions in the Sociology and History of Technology* (anniversary edition), edited by Wiebe E. Bijker, Thomas P. Hughes, and Trevor Pinch, 11–44. Cambridge, MA: MIT Press.

Pitkin, Hanna Fenichel. 1972. *Wittgenstein and Justice: On the Significance of Ludwig Wittgenstein for Social and Political Thought*. Berkeley, CA: University of California Press.

Plato. 1997. "Republic." In *Plato: Complete Works*, edited by John M. Cooper, 971–1223. Indianapolis: Hackett.

Poe, Marshall T. 2011. *A History of Communications: Media and Society from the Evolution of Speech to the Internet*. Cambridge: Cambridge University Press.

Poggi, Gianfranco. 2006. "Political Power Un-Manned: A Defence of the Holy Trinity from Mann's Military Attack." In *An Anatomy of Power: The Social Theory of Michael Mann*, edited by John A. Hall and Ralph Schroeder, 135–149. Cambridge: Cambridge University Press.

Polsby, Nelson W. 1960. "How to Study Community Power: The Pluralist Alternative." *Journal of Politics* 22, 474–484.

Portmess, Lisa, and Sara Tower. 2015. "Data Barns, Ambient Intelligence and Cloud Computing: The Tacit Epistemology and Linguistic Representation of Big Data." *Ethics and Information Technology* 17, no. 1: 1–9.

Prior, Elizabeth W., Robert Pargetter, and Frank Jackson. 1982. "Three Theses about Dispositions." *American Philosophical Quarterly* 19, no. 3: 251–257.

Puente, John G. 2010. "The Emergence of Commercial Digital Satellite Communications." *IEEE Communications Magazine* 48, no. 7: 16–20.

Rabinow, Paul, and Nikolas Rose. 2006. "Biopower Today." *BioSocieties* 1, no. 2: 195–217.

Rabinow, Paul. 1996. *Essays on the Anthropology of Reason*. Princeton, NJ: Princeton University Press.

Raz, Joseph. 1986. *The Morality of Freedom*. Oxford: Clarendon Press.

Raz, Joseph. 1997. "Incommensurability and Agency" In *Incommensurability, Incomparability and Practical Reason*, edited by Ruth Chang, 110–128. Cambridge: Harvard University Press.

Reese, Hope. 2016. "Autonomous Driving Levels 0 to 5: Understanding the Differences." *TechRepublic*, January 20, 2016. https://www.techrepublic.com/article/autonomous-driving-levels-0-to-5-understanding-the-differences/.

Reidenberg, Joel R. 2002. "Yahoo and Democracy on the Internet." *Jurimetrics* 42, 261–280.

Reporters Without Borders. 2009. *Internet Enemies*. https://www.reporter-ohne-grenzen.de/fileadmin/pdf/Internetbericht.pdf.

Ricoeur, Paul. 1986. *Lectures on Ideology and Utopia*. New York: Columbia University Press.

Robinson, Nicholas W., Chen Zeng, and R. Lance Holbert. 2018. "The Stubborn Pervasiveness of Television News in the Digital Age and the Field's Attention to the Medium, 2010–2014." *Journal of Broadcasting & Electronic Media 62*, no. 2: 287–301.

Rose, Nikolas. 2007. *The Politics of Life Itself: Biomedicine, Power, and Subjectivity in the Twenty-First Century*. Princeton, NJ: Princeton University Press.

Rose, Nikolas, and Carlos Novas. 2004. "Biological Citizenship." In *Blackwell Companion to Global Anthropology*, edited by Aihwa Ong, and Stephen J. Collier. S. Oxford: Blackwell.

Rothman, Steven B. 2011. "Revising the Soft Power Concept: What Are the Means and Mechanisms of Soft Power?" *Journal of Political Power* 4, no. 1: 49–64.

Rowlands, Michael. 2005. "A Materialist Approach to Materiality." In *Materiality*, edited by Daniel Miller, 72–87. Durham, NC: Duke University Press.

Rubin, Michael. 2001. "The Telegraph, Espionage, and Cryptology in Nineteenth Century Iran." *Cryptologia* 25, no. 1: 18–36.

Russell, Bertrand. 2004. *Power: A New Social Analysis*. London: Routledge. [First published in 1938 by George Allen & Unwin Ltd, London].

Ryan, Kevin. 2007. *Social Exclusion and the Politics of Order*. Manchester, UK: Manchester University Press.

Ryan, Kevin. 2011. "Exclusion." In *Encyclopedia of Power*, edited by Keith Dowding, 224–228. London: Sage.

Ryle, Gilbert. 2009. *The Concept of Mind* (60th anniversary edition). London: Routledge.

Saar, Martin. 2010. "Power and Critique." *Journal of Power* 3, no. 1: 7–20.

SAE International. 2016. "Taxonomy and Definitions for Terms Related to Driving Automation Systems for On-Road Motor Vehicles." https://www.sae.org/standards/content/j3016_201609.

Saint Augustine of Hippo. 2009 *The City of God*. Translated by Marcus Dods. Peabody, MA: Hendrickson Publishers.

Sandberg, Anders. 2012. "The Censor and the Eavesdropper: The Link between Censorship and Surveillance." Oxford University. *Practical Ethics: Ethics in the News* (blog), March 2, 2012. http://blog.practicalethics.ox.ac.uk/2012/03/the-censor-and-the-eavesdropper-the-link-between-censorship-and-surveillance.

Sattarov, Faridun. 2016. "Coded Power: A Crossdisciplinary Framework for Understanding Algorithmic Power", Liverpool Symposium on Machine Learning: Formulating an Innovation Policy for the 4th Industrial Revolution, July 2016. Liverpool, United Kingdom.

Sattarov, Faridun, and Mark Coeckelbergh. 2017. "Prejudiced Algorithms?: Some Arguments about Algorithmic Power and Bias", CEPE/Ethicomp Conference, June 2017. Turin, Italy.

Sattarov, Faridun, and Saskia Nagel. 2019. "Building Trust in Persuasive Gerontechnology: User-centric and Institution-centric Approaches." *Gerontechnology* 18, 1: 1–14.

Saunders, Nicholas J. 2000. "Bodies of Metal, Shells of Memory: 'Trench Art', and the Great War Re-cycled." *Journal of Material Culture* 5, no. 1: 43–67.

Schattsneider, Eric Elmer. 1960. *The Semi-Sovereign People: A Realist's View of Democracy in America*. New York: Holt, Rinehart & Wilson.

Schmandt-Besserat, Denise. 1992. *Before Writing, Volume I: From Counting to Cuneiform*. Austin: University of Texas Press.

Sclove, Richard. 1992. "The Nuts and Bolts of Democracy: Democratic Theory and Technological Design." In *Democracy in a Technological Society*, edited by Langdon Winner, 139–157. Dordrecht: Springer.

Sclove, Richard. 1995. *Democracy and Technology*. New York: Guilford.

Searle, John R. 2010. *Making the Social World: The Structure of Human Civilization*. Oxford: Oxford University Press.

Seeman, Melvin. 1959. "On the Meaning of Alienation." *American Sociological Review* 24, no. 6: 783–791.

Sen, Amartya. 1999. *Development as Freedom*. New York: Knopf.

Seneca. 2011. *On Benefits*. Translated by Miriam Griffin and Brad Inwood. Chicago: University of Chicago Press.

Sengers, Phoebe, K. Boehner, S. David, and Joseph Kaye. 2005. "Reflective Design." Proceedings of the 4th Decennial Conference on Critical Computing: Between Sense and Sensibility, 49–58. New York: ACM.

Shariff, Azim, Jean-François Bonnefon, and Iyad Rahwan. 2017. "Psychological Roadblocks to the Adoption of Self-Driving Vehicles." *Nature Human Behaviour* 1, no. 10: 694–696.

Shibutani, Tamotsu. 1955. "Reference Groups as Perspectives." *American Journal of Sociology* 60, no. 6: 562–569.

Shively, W. Phillips. 2018. *Power and Choice: An Introduction to Political Science*. London: Rowman & Littlefield.

Simon, Judith. 2010. "The Entanglement of Trust and Knowledge on the Web." *Ethics and Information Technology* 12, no. 4: 343–355.

Simon, Judith. 2015. "Distributed Epistemic Responsibility in Hyperconnected Era." In *The Onlife Manifesto: Being Human in a Hyperconnected Era*, edited by Luciano Floridi, 145–159. Dordrecht: Springer.

Simon, Judith. 2017. "Value-Sensitive Design and Responsible Research and Innovation." In *The Ethics of Technology: Methods and Approaches*, edited by Sven Ove Hansson, 219–235. London: Rowman & Littlefield.

Simons, Jon. 1995. *Foucault and the Political*. London: Routledge.

Singer, Peter. 1973. *Democracy and Disobedience*. Oxford: Oxford University Press.

Smith, E. Stratford. 1970. "The Emergence of CATV: A Look at the Evolution of a Revolution." *Proceedings of the IEEE* 58, no. 7: 967–982.

Smith, Ronald D. 2017. *Strategic Planning for Public Relations* (5th edition). New York: Routledge.

Snake-Beings, Emit. 2013. From Ideology to Algorithm: The Opaque Politics of the Internet. *Transformations: Journal of Media and Culture*, 23.

Solow, Robert M. 1965. "A Contribution to the Theory of Economic Growth." *The Quarterly Journal of Economics* 70, no. 1: 65–94.

Solow, Robert M. 1970. *Growth Theory: An Exposition*. Oxford: Clarendon Press.

Søraker, Johnny, and Philip Brey. 2014. "Deliverable 1.1.: Systematic Review of Industry Relevant RRI Discourses." Responsible-Industry Project. http://www.responsible-industry. eu/dissemination/deliverables.

Spahn, Andreas. 2012. "And Lead Us (Not) into Persuasion . . . ? Persuasive Technology and the Ethics of Communication." *Science and Engineering Ethics* 18, no. 4: 633–650.

Spinoza, Benedict de. 2002. "Political Treatise." In *Spinoza: Complete Works*, edited by Michael L. Morgan, 676–754. Indianapolis: Hackett.

Spinoza, Benedict de. 2002. "Ethics." In Spinoza: Complete Works, edited by Michael L. Morgan, 213–382. Indianapolis: Hackett.

Spinoza, Benedict de. 2002. "Theological-Political Treatise." In *Spinoza: Complete Works*, edited by Michael L. Morgan, 383–583. Indianapolis: Hackett.

Stafford, Ned. 2005. "German Ethics Council under Fire." *The Scientist*, August 1, 2005. https://www.the-scientist.com/news-analysis/german-ethics-council-under-fire-48482.

Stahl, Bernd Carsten. 2012. "Responsible Research and Innovation in Information Systems." *European Journal of Information Systems* 21, 207–211.

Steger, Manfred B. 2007. "Globalization and Ideology." In *The Blackwell Companion to Globalization*, edited by George Ritzer, 367–382. Malden, MA: Blackwell Publishing.

Steiner, George. 1971. "The Mandarin of the Hour—Michel Foucault." Review of *The Order of Things: An Archaeology of the Human Sciences*, by Michel Foucault. February 28, 1971.

Steiner, Hillel. 2007. "Hillel Steiner, from An Essay on Rights (1994)." In *Freedom: A Philosophical Anthology*, edited by Ian Carter, Matthew H. Kramer, and Hillel Steiner, 283–292, Oxford: Blackwell.

Stilgoe, Jack, Richard Owen, and Phil Macnaghten. 2013. "Developing a Framework for Responsible Innovation." *Research Policy* 42, no. 9: 1568–1580.

Strayer, Joseph R. 1973. *On the Medieval Origins of the Modern State*. Princeton, NJ: Princeton University Press.

Sumner, Leonard W. 1996. Welfare, Happiness, and Ethics. Oxford: Clarendon Press.

Sunstein, Cass R. 1986. "Pornography and the First Amendment." *Duke Law Journal*, no. 4: 589–627.

Sweeney, Latanya. 2013. "Discrimination in Online Ad Delivery." *Queue* 11, no.3: 10:10–10:29.

Taebi, Behnam, Aad Correlje, Edwin Cuppen, Marloes Dignum, and Udo Pesch. 2014. "Responsible Innovation as an Endorsement of Public Values: The Need for Interdisciplinary Research." *Journal of Responsible Innovation* 1, no. 1: 118–124.

Takács, Sarolta A. 2008. *The Construction of Authority in Ancient Rome and Byzantium: The Rhetoric of Empire*. Cambridge: Cambridge University Press.

Taylor, Charles. 1984. "Foucault on Freedom and Truth." *Political Theory* 12, no. 2: 152–183.

Taylor, Charles. 1991. *The Ethics of Authenticity*. Cambridge, MA: Harvard University Press.

Taylor, Richard. 1966. *Action and Purpose*. Englewood Cliffs, NJ: Prentice Hall.

Thaler, Richard H., and Cass Sunstein. 2008. *Nudge: Improving Decisions about Health, Wealth and Happiness*. New Haven, CT: Yale University Press.

Thomson, Judith Jarvis. 1976. "Killing, Letting Die, and the Trolley Problem." *The Monist* 59, no. 2: 204–217.

Tilley, Christopher. 1991. *Material Culture and Text: The Art of Ambiguity*. London: Routledge.

Treanor, Jill. 2016. "U.S. Corporate Giants Hoarding More Than a Trillion Dollars." *The Guardian*, May 20, 2016. https://www.theguardian.com/business/2016/may/20/us-corporate-giants-hoarding-over-a-trillion-dollars-apple-microsoft-google.

Tritter, Jonathan Quetzal, and Alison McCallum. 2006. "The Snakes and Ladders of User Involvement: Moving beyond Arnstein." *Health Policy* 76, no. 2: 156–168.

Tucker, Robert C. 1978. *The Marx-Engels Reader* (2nd edition). New York: W.W. Norton.

Uhr, John. 2011. "Persuasion." In *Encyclopedia of Power*, edited by Keith Dowding, 478–479. London: Sage.

Ullmann, Walter. 1965. *A History of Political Thought: The Middle Ages*. Harmondsworth: Penguin Books.

Unger, Stephen H. 1994. *Controlling Technology: Ethics and the Responsible Engineer* (2nd edition). New York: John Wiley & Sons.

van Creveld, Martin. 1991. *Technology and War From 2000 BC to the Present*. New York: The Free Press.

van de Poel, Ibo. 2011. "The Relation between Forward-Looking and Backward-Looking Responsibility." In *Moral Responsibility: Beyond Free Will and Determinism*, edited by Nicole A. Vincent, Ibo van de Poel, and Jeroen van den Hoven, 37–52. Dordrecht: Springer.

van Dijk, Jan A. G. M. 2006. *The Network Society*. London: Sage.

van Oudheusden, Michiel. 2014. "Where Are the Politics in Responsible Innovation? European Governance, Technology Assessments, and Beyond." *Journal of Responsible Innovation* 1, no. 1: 67–86.

Varga, Somogy, and Charles Guignon. 2017. "Authenticity." *The Stanford Encyclopedia of Philosophy*, edited by Edward N. Zalta. https://plato.stanford.edu/archives/fall2017/entries/authenticity.

Vaughan, Liwen, and Mike Thelwall. 2004. "Search Engine Coverage Bias: Evidence and Possible Causes." *Information Processing & Management* 40, no. 4: 693–707.

von Schomberg, René. 2012. "Prospects for Technology Assessment in a Framework of Responsible Research and Innovation." In *Technikfolgen Abschätzen Lehren*, edited by Marc Dusseldorp and Richard Beecroft, 39–61. Springer Science & Business Media.

von Schomberg, René. 2013 "A Vision of Responsible Research and Innovation." In *Responsible Innovation: Managing the Responsible Emergence of Science and Innovation in Society*, edited by Owen, Richard, John R. Bessant, and Maggy Heintz, 51–74. London: John Wiley & Sons.

von Wright, Georg Henrik. 1963. *Norm and Action: A Logical Enquiry*. London: Routledge & Kegan Paul.

Wachter, Sandra, Brent Mittelstadt, and Chris Russell. 2018. "Counterfactual Explanations without Opening the Black Box: Automated Decisions and the GDPR." *Harvard Journal of Law & Technology* 31, no. 2.

Waddell, Paul, Alan Borning, Michael Noth, Nathan Freier, Michael Becke, and Gudmundur Ulfarsson. 2003. "Microsimulation of Urban Development and Location Choices: Design and Implementation of UrbanSim." *Networks and Spatial Economics* 3, no. 1: 43–67.

Wajcman, Judy. 1991. *Feminism Confronts Technology*. Cambridge, UK: Polity Press.

Waldron, Jeremy. 2002. "Is the Rule of Law an Essentially Contested Concept (in Florida)?" *Law and Philosophy* 21, no. 2: 137–164.

Wallach, Wendell, and Colin Allen. 2010. *Moral Machines: Teaching Robots Right from Wrong*. Oxford: Oxford University Press.

Warren, Mark. 1988. *Nietzsche and Political Thought*. Cambridge, MA: MIT Press.

Wartenberg, Thomas, 1990. *The Forms of Power: From Domination to Transformation*. Philadelphia: Temple University Press.

Weber, Max. 1946. *Essays on Sociology*. New York: Oxford University Press.

Weber, Max. 1947. *Max Weber: The Theory of Social and Economic Organization*. Translated by A. M. Henderson, edited by Talcott Parsons. New York: The Free Press.

Webster, Frank. 1995. *Theories of the Information Society*. London: Routledge.

Weedon, Chris. 1987. *Feminist Practice and Poststructuralist Theory*. Oxford: Blackwell.

White, Lynn Jr. 1964. *Medieval Technology and Social Change*. Oxford: Oxford University Press.

White, Stephen. 1986. "Economic Performance and Communist Legitimacy." *World Politics* 38, no. 3: 462–482.

Wickham, G. 1983. "Power and Power Analysis." *Economy and Society* 12, no. 4: 468–498.

Williams, Garrath. 2008. "Responsibility as a Virtue." *Ethical Theory and Moral Practice* 11, 455–470.

Winner, Langdon. 1980. "Do Artifacts Have Politics?" *Daedalus* 109, no. 1: 121–136.

Winner, Langdon. 1986. *The Whale and the Reactor*. Chicago: University of Chicago Press.

Winner, Langdon. 1993. "Upon Opening the Black Box and Finding It Empty: Social Constructivism and the Philosophy of Technology." *Science, Technology & Human Values* 18: 362–378.

Winner, Langdon. 1995. "Citizen Virtues in a Technological Order." In *Technology and the Politics of Knowledge*, edited by Andrew Feenberg and Alastair Hannay, 65–85. Bloomington, IN: Indiana University Press.

Wittgenstein, Ludwig. 1967. *Philosophical Investigations*. Oxford: Oxford University Press.

Wolin, Sheldon S. 2004. *Politics and Vision: Continuity and Innovation in Western Political Thought* (2nd expanded edition). Princeton, NJ: Princeton University Press.

Wrong, Dennis. 1968. "Some Problems in Defining Social Power." *American Journal of Sociology* 73, no. 6: 673–681.

Wu, Tim. 2010. *The Master Switch: The Rise and Fall of Information Empires*. London: Atlantic Books.

Yeung, Karen. 2017. "'Hypernudge': Big Data as a Mode of Regulation by Design." *Information, Communication & Society* 20, no. 1: 118–136.

Young, Iris Marion, 1990. *Justice and the Politics of Difference*. Princeton, NJ: Princeton University Press.

Zaloom, Caitlin. 2006. *Out of the Pits: Traders and Technology from Chicago to London*. Chicago: University of Chicago Press.

Zimmermann, Andrew D. 1995. "Toward a More Democratic Ethic of Technological Governance." *Science, Technology & Human Values* 20, no. 1: 86–107.

Žižek, Slavoj. 1989. *The Sublime Object of Ideology*. London: Verso.

Zwart, Hub, Laurens Landeweerd, and Arjan van Rooij. 2014. "Adapt or Perish? Assessing the Recent Shift in the European Research Funding Arena from 'ELSA' to 'RRI'." *Life Sciences, Society and Policy* 10, http://www.lsspjournal.com/content/10/1/11.

·

Index

ability, 5, 19, 20, 45–47, 49, 50, 101, 118, 121, 123
Acemoglu, Daron, 154–157, 163, 164
actant (nonhuman), 37, 54, 77
actantional, 54. *See also* actant (non-human)
advertising, 36, 67, 103
affordance, 4, 53, 78
agency, 51–52; distributed, 121; moral, 49; technological, 52–55
algorithmisation, 104
algorithms, 4; definitions of, 99; ethics of, 8, 10, 97, 115; and high-frequency trading (HFT), 104, 105; machine-learning, 50, 97, 109–110; moral, 149, 151, 152; and power, 99–105
alibi, 118
alienation, 15, 77
Al-Khwārizmī, 99
Allen, Amy, 6, 22, 23
Apple, 72, 79
Arendt, Hannah, 5, 14, 19, 23, 123
Aristotle, 8, 43, 44–45, 49, 58, 101, 126
army, 41, 59, 85
Arnstein, Sherry, 141
Augustine, 73
Austin, John, 43, 45–46
authenticity, 9, 10, 117, 125, 128–129, 134. *See also* values, authentic
authority, 2, 18, 27, 29, 33, 39, 40; charismatic, 33; competing, 76–78;

delegation of, 39; final, 76, 77, 167; legal, 33; state, 39, 75–76, 84; and technology, 39; traditional, 33
automation, 2, 74, 149
autonomous car. *See* car, autonomous
autonomy, 15; legal, 60, 63; moral, 127; personal, 125, 129

Bachrach, Peter, 18, 28, 29, 30, 31–32, 68
Baier, Annette, 129, 130, 133
Baratz, Morton. *See* Bachrach, Peter
Barnes, Barry, 131, 132–133
Bauman, Zygmunt, 83, 112
Bentham, Jeremy, 8, 86
bias (in computer algorithms), 98; by consequence, 108, 110; informational, 107–108; mobilisation of, 111–112; by nature, 108–109; by nurture, 108, 109–110; in search engines, 108, 109; social, 106; user, 107
Bierstedt, Robert, 31
biocapital, 93
bioeconomics, 90
bioethics, 144
biology, 90
biopolitics. *See* biopower
biopower, 90–92
BMVI. *See* German Federal Ministry of Transport and Digital Infrastructure
Bok, Sissela, 129
bow (and arrow), 73

About the Author

Faridun Sattarov received his PhD in Philosophy of Technology from the University of Twente. He has been a research fellow at the UNESCO Bioethics and Ethics of Science Section, University of Liverpool School of Law, Technological University of Eindhoven and University of Twente. He is an affiliate member of the 4TU.Centre for Ethics of Technology.

www.ingramcontent.com/pod-product-compliance
Lightning Source LLC
Chambersburg PA
CBHW021816270326

41932CB00007B/216